W9-CHF-333

The Ecological Basis for

JAMES L. NEWMAN

Subsistence Change among the Sandawe of Tanzania

FOREIGN FIELD RESEARCH PROGRAM

NATIONAL ACADEMY OF SCIENCES
WASHINGTON, D.C. 1970

ISBN 0-309-01851-X

Available from

Printing and Publishing Office
National Academy of Sciences
2101 Constitution Avenue
Washington, D.C.   20418

Library of Congress Catalog Card Number 75-607171

Printed in the United States of America

PREFACE

The initial plan for this study called for a comparative analysis of subsistence change among the Sandawe and Hadza (Kindiga) peoples of Tanzania. Standard ethnographic sources for East Africa have generally linked the two groups culturally and historically, though the Hadza follow a hunting and gathering way of life while the Sandawe have adopted a crop and animal husbandry economy. An obvious question, then, might be, Why had the one changed while the other had not? As this study progressed, it became apparent that this would be a very difficult, and perhaps impossible, question to answer. Further research into published and unpublished materials and conversations with knowledgeable investigators convinced me that there is no clear evidence that the Sandawe and Hadza ever shared a common cultural tradition. In fact, as we shall see, the evidence suggests that they are in no way related. It thus appeared most profitable to focus on the Sandawe—to examine with some care the view that they had evolved from nomadic hunters and gatherers into settled husbandmen.

At the completion of a preliminary field survey, the conviction grew that merely contributing to the solution of a historical problem might not be quite enough. In a country embroiled in the difficult and painful process of profound socioeconomic change, could a useful contribution be made as well? The decision to try led to the inclusion of data related to possible future changes, and thus the final form of the study emerged.

Field research was carried out between March 1965 and May 1966 under a grant administered by the Foreign Field Research Program, Division of Earth Sciences, National Academy of Sciences–National Research Council. Except for four weeks in London and occasional visits to Dar es Salaam, Nairobi, and Kampala, almost all of this period was spent in Sandawe country.
for four weeks in London and occasional visits to Dar es Salaam, Nairobi, and Kampala, almost all of this period was spent in Sandawe country.

ACKNOWLEDGMENTS

To the Sandawe, I express heartfelt gratitude. Their willingness to accept a nosy stranger and to share their hospitality and lore with him was beyond all my expectations. I doubt whether a more congenial atmosphere exists anywhere. I particularly want to single out Longini Tamba, my erstwhile assistant. The many Tanzanian officials, especially in the Ministry of Agriculture, who generously gave time and information and helped smooth over administrative difficulties are also acknowledged and thanked.

A very special debt of gratitude is owed to Philip W. Porter for his patient advice, counsel, and criticism from the formulation of the study through its completion. W.F.E.R. Tenraa, whose name you will see repeatedly in the text, cannot be thanked enough. His help with the Sandawe language and his gracious sharing of personally collected data have added much that would otherwise be missing. To Miss Judy Norman for her drawings and to Mrs. Patricia Burwell for her cartography I offer sincere appreciation. Finally, there is my wife, Carole. She has been through it all, from start to finish, and her constant encouragement and assistance, which never faltered despite a serious accident and an attack of malaria, provided the most valuable incentive.

CONTENTS

I. INTRODUCTION

1. STRATEGY AND ORGANIZATION 3

2. GENERAL BACKGROUND 6
Physical Framework, 6
Human Framework, 17

II. CHANGING SANDAWE SUBSISTENCE PATTERNS

3. THE HUNTING AND GATHERING PAST 25
Oral Tradition and Literature, 25
The Role of Hunting and Gathering, 27
The Underdeveloped Husbandry Tradition, 35
Other Indicators, 43

4. THE ADOPTION OF CROP AND ANIMAL HUSBANDRY 47

5. RECENT CHANGE 57

III. SUBSISTENCE POTENTIAL

6. CROP PRODUCTION 69
Water Balance, 70
Occurrences of Famine, 76
Surface-Water and Groundwater Resources, 80
Edaphic Conditions, 82
Diseases and Pests, 102

7. LIVESTOCK PRODUCTION 129
Practices and Problems, 129
Tsetse and Trypanosomiasis, 136

8. OTHER MEANS OF SUBSISTENCE 141
Game and Hunting, 141
Bush Products, 143
Fish, 149

9. POPULATION-LAND BALANCE 151

IV. CONCLUSION

10. CONCLUSION 159

APPENDIXES

APPENDIX A: SELECTED ITEMS OF SANDAWE MATERIAL CULTURE 165

APPENDIX B: ANIMALS AND BIRDS HUNTED BY THE SANDAWE 171

APPENDIX C: PLANTS USED BY THE SANDAWE 176

NOTES 185

REFERENCES 193

I

INTRODUCTION

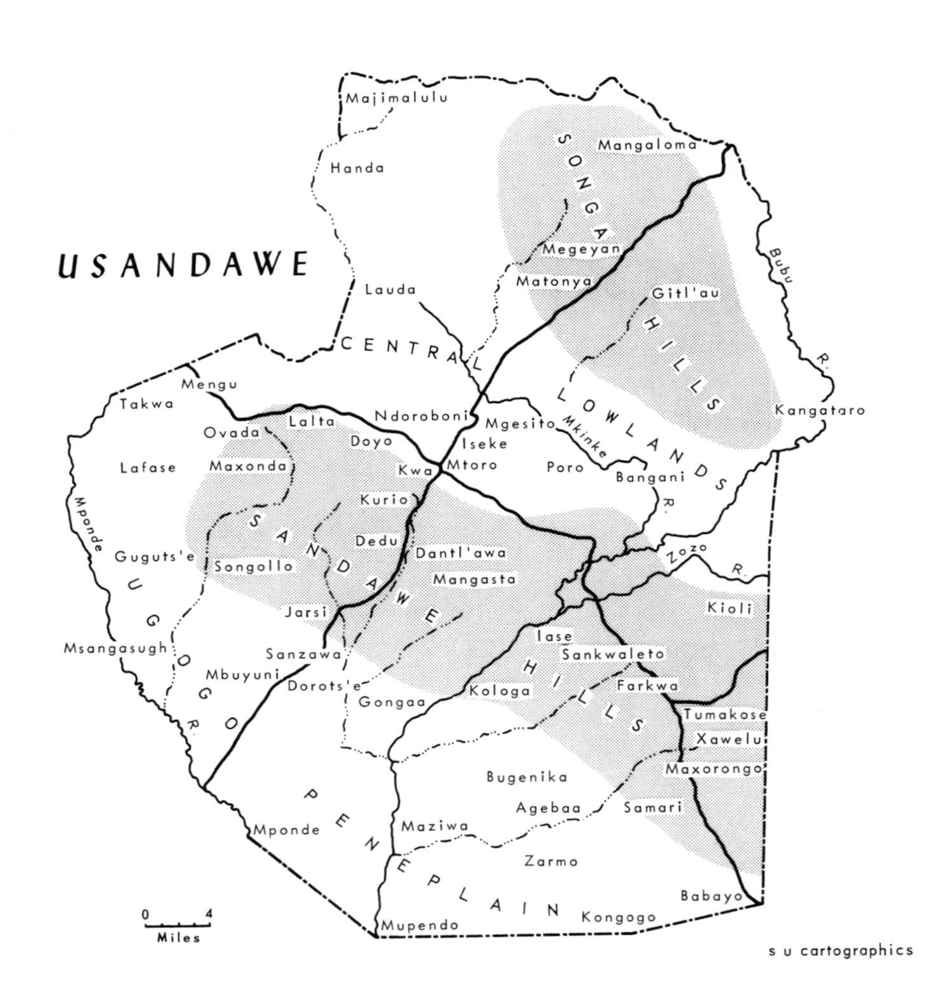
USANDAWE
Majimalulu
Handa
Mangaloma
SONGA HILLS
Megeyan
Matonya
Gitl'au
Bubu R.
Lauda
CENTRAL LOWLANDS
Kangataro
Mengu
Takwa
Lalta
Ndoroboni
Mgesito
Mkinke R.
Ovada
Doyo
Iseke
Lafase
Maxonda
Kwa
Mtoro
Poro
Bangani
Kurio
Mponde
SANDAWE HILLS
Dedu
Zozo R.
Guguts'e
Songollo
Dantl'awa
Mangasta
UGOGO
Kioli
Jarsi
Iase
Msangasugh
Sanzawa
Sankwaleto
Mbuyuni
Dorots'e
Farkwa
Gongaa
Kologa
R.
Tumakose
Xawelu
Maxorongo
Bugenika
Agebaa
Samari
PENEPLAIN
Mponde
Maziwa
Zarmo
Babayo
0
4
Miles
Mupendo
Kongogo
s u cartographics

CHAPTER 1

STRATEGY AND ORGANIZATION

This study of the changing subsistence modes of the Sandawe people of central Tanzania follows Ackerman in the assumption that change has a spatial dimension.[1] That is, the form and intensity of change are place-related and therefore susceptible to a geographic approach. Within a spatial framework, both macroscopic and microscopic perspectives are possible. On the macroscopic plane, the processes and phenomena of change are viewed as aggregates of discrete cases. The essential aim is to establish broadly based generalizations that will lead to the formulation of high-level theory. In consequence, much detail is discarded along the way as "noise." Good examples of such an approach are the innovation wave studies of Hägerstrand and his followers.[2] A microscopic orientation, the one followed in the present work, is virtually synonymous with what is now termed ecology. Universality is sacrificed for inclusiveness and detail, the goal being to uncover the functional interrelations of the physical and behavioral features of the specific cases that underlie the broader patterns. As Blaut has indicated, it is usually with data in this form that geographers can best inform and influence policy makers.[3]

Ecology is used here in its most widely accepted sense, namely that of the interaction between community and environment. Because of the confusion and ambiguity that generally have accompanied the extension of ecology from biology to the social sciences, both terms require brief elaboration. To the biological ecologist, communities are composed of a "definable species content, a certain quality of stability, and a clear organization."[4] Using these criteria, many human communities of differing size and composition could be demarcated, but I suggest that throughout most of rural Africa, and certainly Tanzania, the commonly recognized ethnic groups (tribes) are the most useful and relevant entities. Their members can be fairly easily identified; they persist by propagating common norms and values, and they provide the main force behind institutional organization. The national state systems have not yet supplanted them as the largest collectivities to which people readily subordinate their actions.

Usually the term "environment" is used to connote the physical or natural universe. However, the literal interpretation is "surroundings," or all factors impinging upon a subject. Consequently, when we speak of human communities, it is necessary to recognize that they react upon and influence one another as well as interacting with the nonhuman, or as Sahlins has put it, "societies are typically set in fields of *cultural influence* as well as fields of natural influence."[5] The contact situation, then, becomes part of the environment of ecology.

In interpreting the interaction between community and environment, a prominent position must be given to cultural cognition. Based on its own experience, every society establishes criteria and standards for evaluating phenomena, and therefore it is not sufficient to plug observed data into the preconceived categories of Western science. Numerous schemes for change throughout the world have demonstrated aptly that these categories are often both incomplete and inadequate. One must "also describe the environment as the people themselves construe it according to the categories of their own ethnoscience."[6] Only in this way is it possible to ascertain the true meaning of the environment to the functioning of social institutions and behavior.

Another assumption of the present study is that change should be looked at in a historical context. The notion of static societies, even those with the most rudimentary technological equipment, has long been dispelled. Every change is grafted on to an already existing process of "becoming." The future is not disjoint from the past, a fact to which social planners might well pay more attention.

In line with the above discussion, there are four parts to this presentation. The remainder of Part I is devoted to a discussion of the general physical environmental and cultural setting of the Sandawe. Part II attempts to trace past changes in Sandawe subsistence practices in order to uncover what forces have been traditionally operative. Particular attention is given to the reported transition the Sandawe have made from a nomadic hunting and gathering existence to one of basically settled husbandry. With this background for context and direction, Part III attempts to assess the resource potential of the environment. Part IV presents the author's major conclusions.

Before proceeding, it should be re-emphasized clearly that we are concerned here with agriculture and other related subsistence activities. The habitat of the Sandawe apparently lacks valuable minerals in any workable quantity,[7] and no industrial-commercial enterprises are foreseen in the near future. As is the pattern over so much of Africa, and in the third world generally, the agrarian sector bears the brunt of economic growth, often being "the only base . . . from which to initiate general economic developments."[8] According to Allan, the demands placed upon agriculture require it to

> 1. Generate capital for the economic transition, including foreign exchange gained by land products;
> 2. Provide the labour force required for expansion of the non-agricultural sectors of the economy;
> 3. Meet the growth in demand for agricultural products created by industrial consumption, and increasing non-rural population, and rising per capita incomes; and
> 4. Support a surplus population until such time as this can be absorbed by the general expansion of the economy.[9]

He therefore concludes that agriculture "has the heavy task of creating capital both for investment in other sectors and for its own transformation.[10]

To sum up briefly, this is an ecologically oriented study of the course of subsistence change among the Sandawe of central Tanzania. It first tries to examine the internal and the external, the social and the physical forces that have produced the patterns we now see. It then goes on to provide a base from which future change and development can be planned and understood.

CHAPTER 2

GENERAL BACKGROUND

PHYSICAL FRAMEWORK

Sandawe country, or Usandawe,[1] is situated on the southern margins of what is widely known as the Central Highlands.[2] In the present-day administrative framework, this larger area closely approximates the Mbulu, Iramba, Singida, and Kondoa districts (Map 1). The terrain varies from gently undulating plateau country to extremely flat plains, all of which have been greatly influenced by the faulting of the Gregory rift system (Map 2). There are several prominent fault escarpments, and both Lake Eyasi and Lake Manyara owe their existences to the process of rifting. The elevation is nearly everywhere over 3,000 ft, with a considerable portion above 5,000 ft.

The climate is typically semiarid. Excluding the heights over 7,000 ft in the north, rainfall averages under 30 in. per year and is characterized by marked variability. Usandawe has two recording stations: one at Kurio, the other at Farkwa. Through 1965, the former had been in operation for 42 years and had recorded an annual average of 25.1 in., while the latter, over a 23-year span from 1943, gave an average of 23.8 in. Rainfall at Farkwa has ranged from a low of 10.9 in. to a high of 36.7 in. In consecutive years, Kurio records went from 14.4 in. to 40.4 in, its lowest and highest readings. Precipitation is concentrated in one season, from

MAP 1 Administrative districts of Tanzania.

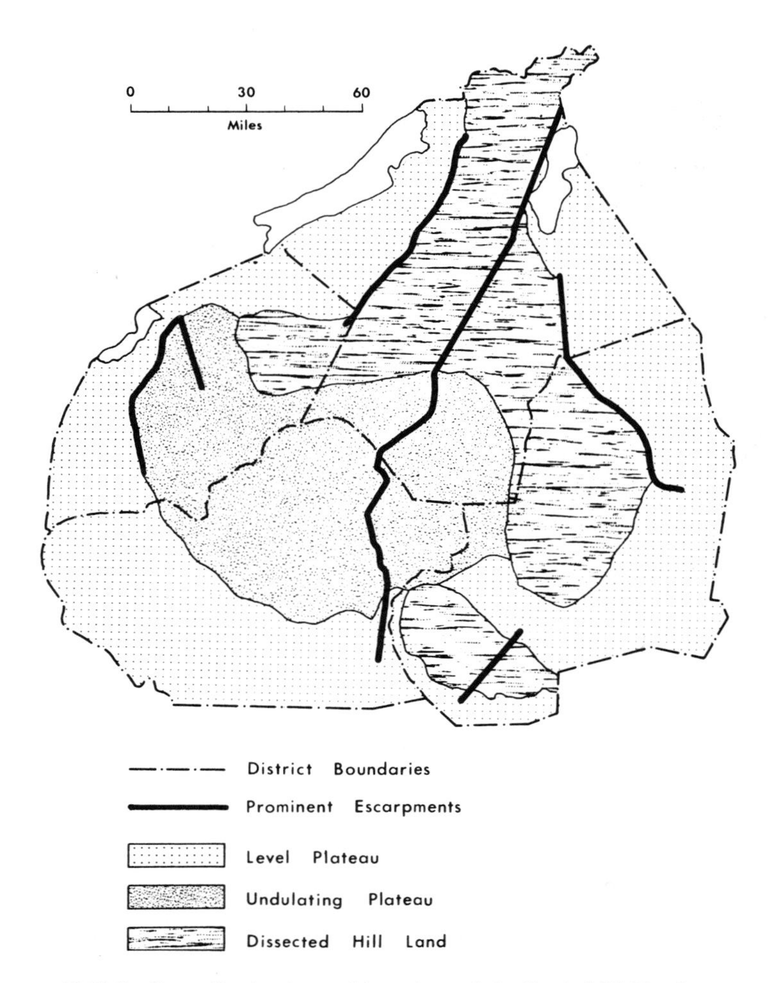

MAP 2 Generalized topographic regions of the Central Highlands.

late November through April, when some 95 percent of the yearly total falls. Almost invariably, however, there is a dry spell of at least several weeks duration sometime after the initial rains have begun. This dry spell, in conjunction with the factor of variability, gives rise to periodic severe food shortages throughout the Central Highlands. More is said about the problems of moisture resources and famine later in the report.

The geology, physiography, vegetation, and soils of Usandawe all present a varied and complexly interwoven pattern. The "heartland" consists of a series of northwest-southeast trending ridges called the Sandawe Hills (Figure 1). The crests—excluding a few old remnants that project above the general level, the most notable being Doyo with an elevation of 5,608 ft—stand at fairly even heights of from 4,500–4,700 ft. The valley bottoms are generally at elevations between 3,700 and 3,800 ft. According to Gillman, these uplands, along with the Chenene Hills in Dodoma district just to the southeast, constitute a product of fairly recent posttectonic dissection, probably Plio-Pleistocene.[3] Numerous fault lines cut through the hills, most of them running transversally in a northeast-southwest direction. The Bubu fault bisects the Sandawe Hills, providing a course for the principal river of the Central Highlands, the Bubu. An almost vertical escarpment rises sharply on the right bank of the stream (Figure 2). In the west, the Mponde fault controls the course of the Mponde River and is associated with an even more imposing escarpment, which, however, is located outside of Usandawe in Singida district (Figure 3). Many watercourses that are dry except for brief periods following heavy rainstorms are also located along fault lines, but no other noticeable escarpments are encountered.

FIGURE 1 A view westward across the Sandawe Hills near Farkwa. It is early June, the dry season has just begun, and thus the vegetation cover is still full.

FIGURE 2 The Bubu escarpment marks the right bank of the Bubu River as it traverses the Sandawe Hills. A recently harvested grain field awaits the grazing of livestock.

FIGURE 3 The Mponde escarpment borders the Mponde River and lies just west of Usandawe in Turu country. Stretching away to the river is a dense stand of *Acacia-Commiphora* woodland and thicket.

The slopes of the Sandawe Hills are mantled with residual sandy to loamy-sand soils that are predominantly dusky red in color, but gray in some places, and that, over their greater part, support a cover of *Brachystegia* or *miombo* woodland.[4] This is the open deciduous type of woodland that, with only minor changes in specific composition, characterizes so much of the vegetational landscape in the semiarid portions of Africa from Tanzania southward (Figures 4, 5). The overwhelmingly dominant species in the Sandawe Hills is *Brachystegia spiciformis.* When in full leaf, it forms a fairly continuous canopy about 40 to 50 ft high. On some of the steeper slopes, the slightly taller *Brachystegia microphylla* is encountered. Other large trees include *Sclerocarya birrea, Pterocarpus*

FIGURE 4 This overview of the *Brachystegia* woodland in the Songa Hills during the height of the dry season shows a landscape virtually devoid of greenery. The only exceptions are a few *Euphorbia* and *Commiphora* species.

FIGURE 5 A Sandawe woman is herding livestock in *Brachystegia* woodland.

angloensis, Entandophragma bussei, and *Terminalia sericea.* An understory having trees of lesser stature—from 10 to 30 ft—is also present. The main species of these smaller trees are *Canthium burttii, Commiphora eminii, Commiphora caerulea, Euphorbia cuneata, Markhamia obtusifolia, Vangueria tomentosa, Fagara chalybea,* and *Holarrhena febrifugia.* There are few low shrubs and bushes.

Valley bottoms, drainage lines, and level areas within the hills (in other words, those sites receiving rainwash from above) have soils of a colluvial nature, mainly gray to brown clay loams.[5] A hardpan layer is typically located within a foot or two of the surface. The vegetation is much denser here than in the woodland, forming a virtual thicket from 12 to 15 ft high in spots where various heavy-thorned *Commiphora* species, particularly *C. ugogensis* and *C. madagascarensis,* interlace with *Cassia abreviata, Lannea humilis,* and *Bussea massaiensis.* A heavy ground cover, up to several feet in height, is provided by *Disperma crenatus* and/or *Barleria taintensis.* Some *Acacia* trees, mainly *A. kirkii, A. tortillis, A. nigrescens,* and very occasionally *A. rovumae,* rise above the general level of the thicket where it is not so dense. Baobabs (*Adansonia digitata*), *Euphorbia bilocularis,* and *Delonix elata* are other occasional members. There is no generally accepted or widely used term for this vegetation assemblage. Trapnell and Langdale-Brown call it *Commiphora* Thicket and Bushland, but the presence of fairly numerous *Acacia* trees indicates that a modifier such as *Acacia-Commiphora* Woodland and Thicket might be a better description.[6]

The ridge crests combine bare granitic outcroppings with pockets of thin stony soils. These latter contain a jumble and tangle of small, essentially nonthorny shrubs, often so densely clustered that they are impossible to move through except on one's hands and knees. *Abrus schimperi, Burttia prunoides, Vitex payos, Grewia Platyclada,* and *Grewia mollis* are the most common members. The only major thorned species is the treacherous *Acacia brevispica.* A few trees, such as *Dalbergia arbutifolia, Albizzia tanganyicensis, Commiphora eminii,* and *Euphorbia bilocularis,* are conspicuous by their greater height. Because of the limited areal extent of this association it is not usually given a separate designation in vegetation classification. However, as we shall see, these sites do have significance with respect to Sandawe agricultural practices, and thus for the sake of clarity and convenience they will be referred to as the Ridge Crest association.

On the south, the Sandawe Hills give way to the vast peneplain of central Ugogo. The lowest elevations in Usandawe are attained in this section, reaching down to 3,000 ft along the Bubu River. The portion roughly east of the 3,200-ft contour line situated on the west side of the Bubu is underlain by colluvial materials and soils similar to those found in the valley bottoms, etc., described above. In like manner, the vegetation is much the same, being characterized by the *Acacia-Commiphora* complex, but a few differences do exist. Baobabs are more frequent as is *Acacia rovumae.* Somewhat greater variety is found among the low shrubs, where *Ximenia americana, Lantana viburnoides, Asparagus racemosus, Disperma crenatum,* and *Commiphora pteleifolia* enter prominently into the picture.

In the extreme southwest and pushing up the Mponde Valley is the northernmost outlier of a geological formation known as the Kilimatinde cement. This is a silcrete, dominated by sandstone and arkose and apparently formed in a shallow basin under contributions from alluvial fans, sand river alluvium, and shallow lake deposits.[7] The concealed Sanzawa fault separates it from the Sandawe Hills, indicating that the cement was formed below an escarpment and antecedent to the faulting. The associated soils are very fine-grained dusky orange to gray silts. However, the unique and most fascinating feature of the area is the vegetation: a dense, in places impenetrable, deciduous thicket known widely as the Itigi thicket (Figure 6). It was first commented on by Burton in his mid-nineteenth century expedition to the Great Lakes and has attracted the attention of numerous observers ever since because of its unique physiognomy and species composition.[8] Milne has termed it a "dwarf forest,"[9] but the most vivid and comprehensive description comes from Burtt. According to him:

> This remarkable vegetation type . . . is composed almost entirely of various species of much branched, coppice forming shrubs, growing from 8 to 15 feet in height. The shrubs are evenly spaced, giving on the whole an open appearance when viewed from inside the thicket; while the shrubs themselves are interlaced overhead to form a thick, even a continuous canopy that becomes very dense in full leaf flush, and shows amazing uniformity when viewed from an aeroplane. The canopy may be pierced by flat topped trees 25 feet high of *Albizzia brachycalyx,* while it is especially in the remoter parts of the thicket, by evergreen woods of *Craibia burttii* 20–35 feet in height.[10]

The most common shrubs in the Usandawe portion of the thicket are *Pseudoprosopis fischeri* and *Combretum trothae,* followed by *Burttia prunoides, Grewia burttii, Bussea massaiensis,* and *Abrus schimperi.*[11]

A similar deciduous thicket is found on the slopes of the Sandawe Hills, interrupting the stands of *Brachystegia* woodland. It is most fully devel-

FIGURE 6 The Itigi thicket as viewed from along the road near Mbuyuni.

oped in the southeast, especially from near Sankwaleto to Maxorongo, but scattered patches are seen virtually throughout the hills. The tendency is for it to be less closed than the Itigi thicket, with a slightly different species composition as well. The dominant shrub is *Justicia salvioides,* joined by lesser numbers of *Combretum trothae, Grewia holstii, Burttia prunoides, Abrus schimperi,* and *Lannea flocussa.* I did not observe any *Pseudoprosopis fischeri* or *Baphia massaiensis.* Trees are more noticeable, including *Combretum apiculatum* and *Ochna ovata,* which rise up to about 15 ft, and *Albizzia petersiana* and *Lonchocarpus eriocalyx,* penetrating to heights of 30 ft.

Just to the north of the Sandawe Hills stretches another band of quite level country, here with elevations of from 3,600 to 4,000 ft. Again, the hardpan clay loam soils derived from colluvial materials, with the associated *Acacia-Commiphora* vegetation, characterize the physical landscape (Figure 7). *Acacia rovumae* are quite numerous and form a striking feature as they reach up to 60 or 70 feet, while baobabs also attain their greatest density. At the same time, the tangle of shrubs and small trees are found to be even more complexly and intensely interwoven than in the other areas. *Commiphora swynnertonii, Grewia platyclada, Grewia mollis,* and *Euphorbia cuneata* join the list of common plants, and very often the only routes of penetration are along the fairly numerous, crisscrossing game trails. Interrupting the *Acacia-Commiphora* pattern in many places are stretches of what are known in East Africa as *mbuga,*[12] or following Gillman's terminology, Valley Grassland.[13] These are extremely flat, open, or sparsely tree-covered grasslands developed on heavy, black clay soils. Surface cracking to a few inches is common during the dry season, whereas in the rains

FIGURE 7 In the lowlands between the Sandawe Hills and the Songa Hills the *Acacia-Commiphora* woodland and thicket can become virtually impenetrable. Seasonal swamps provide one of the few openings.

the topsoil becomes extremely sticky, and travel, except on foot, is impossible. The grasses are mainly various species of *Aristida,* while virtually the only tree is *Acacia drepanolobium,* one of the fascinating ant-gall acacias (Figure 8). A few low and widely scattered *Albizzia anthelmintica* and *Capparis sepiaria* are also encountered.

According to Coster, there are two types of *mbugas* in central Tanzania: those connected with block-faulting where hollows created have been filled by clays and limestones and those in areas of old, sluggish drainage systems connected with widespread peneplanation.[14] The latter would seem to account for most of the mbugas in Usandawe, since it is quite likely that the belt of central lowlands is part of the Ugogo peneplain, having only been separated from it in recent times by the rise of the Sandawe Hills. The *mguba* along the Humbu//han watercourse is probably the result of faulting.[15]

In the northeast rise the Songa Hills. They are, like the Sandawe Hills, of faulted origin and granitic composition, but tend to be slightly higher and more rugged. Ridge elevations are almost universally over 5,000 ft and reach a maximum of 5,707 ft in the peak known as Gitl'au, while valley bottoms are run from about 4,000 to 4,200 ft. The light sandy soils of the hill slopes support an almost exclusive cover of *Brachystegia* woodland, dominated once more by *B. spiciformis.* The *Acacia-Commiphora* association, again, forms ribbons and patches along the drainage lines and bottoms and also characterizes the flat land on the eastern side of the hills leading up to the Bubu River. Unfortunately, I was not able to carry out a careful analysis of species composition in the Songa Hills,

FIGURE 8 An *mbuga* near Kwa Mtoro, showing a stand of ant-gall acacias (*Acacia drepanolobium*).

and, since no previous study is available, more detail will have to await further inquiry.

The far northwest of Usandawe, that is, north of the central lowlands or approximately the 4,000-ft contour line, is characterized by a continuous but fairly gentle rise in elevation, a pattern that continues into Mbulu district and culminates in the Mangatti plains around the slopes of Mount Henang. The soil and vegetation makeup is extremely complex, as a little *Acacia-Commiphora* on localized heavier soils alternates with deciduous thicket of the *Justicia salvioides* variety, *Brachystegia* woodland, and some *Combretum-Terminalia* woodland on the lighter soils. The latter association is found scattered throughout Usandawe as a transition type between *Brachystegia* woodland and the other primary communities. It is also an open, deciduous woodland, but the distinguishing species are now *Combretum zeyheri, Terminalia sericea,* and *Ostryodderis stuhlmannii.* The only place where it reaches any appreciable and coherent extent is where the Sandawe Hills *Brachystegia* gives way to the Itigi thicket.

Along and immediately adjacent to the rivers and watercourses, several interesting vegetation patterns have been produced. Throughout most of the Bubu's course through Usandawe, its immediate channel sides are lined with an evergreen forest of *Kigelia aethiopica, Tamarindus indica,* and *Combretum schumannii.* In areas where the river regularly overflows its banks and floods during the rains, a sparse grassland with scattered *Acacia mellifera* is normally encountered. The latter situation is especially well developed along the Zozo and Mkinke rivers, both of which are quite shallow and sluggish and therefore back up and flood when the Bubu is in full spate. In the northeast where it enters Usandawe, the Bubu has no single discernible channel, but instead spreads out to form a papyrus swamp known as Lake Serya. Extensive flooding is the rule during the rains and some standing water usually remains throughout the dry season. The Mponde River course alternates between sparse grassland where floods are common and deciduous thicket where the channel banks are steep enough to effectively circumscribe the water. Many of the small watercourses are marked by the waving palm fronds of *Hyphaene thebaica.*

One further observation should be made. Up to now, we have been discussing vegetation and soils without reference to the influences of man. Though man has in the past undoubtedly played a role in shaping each of the various associations and is at present actively engaged in modifying them, when viewed in the aggregate he does not appear as an overly prominent agent. However, there is one area in the Sandawe Hills, running roughly in a triangular form between Kwa Mtoro, Ovada, and Dedu, where there seems little question that it has been essentially man who has created and maintained a specific and unique pattern. Here, permanent agricultural fields interspersed with large trees, principally baobabs, *Afzelia quanzensis,* and *Acacia tortilis*, dominate the scene. Between these fields are grazing lands, covered by *Brachiaria, Panicum,* and *Tragus* species of grass and very numerous small herbs. The shrub *Cassia singueana* is also widely present. Where the grazing has been exceptionally heavy and the vegetation cover denuded, the surface of the ground becomes covered by a hard pericap that leads either to gully erosion or to the growth of thick stands of heavy-thorned *Dichrostachys cinerea,* depending on the angle of slope. Because of their open appearance, such areas have traditionally been termed "cultivation steppes."[16] This term has been widely criticized in recent years as an improper usage of the word steppe, and Gillman, who at one time employed it, has suggested that "Actively Induced Vegetation" be substituted in its stead.[17]

Map 3 summarizes the pattern of vegetation distribution; a similar map for soils appears in Chapter 6.

The main parameters of the physical environment have been presented in the preceding discussion as a general guide and orientation to the country. Their significance for changing Sandawe subsistence practices is elaborated on in the forthcoming pages.

HUMAN FRAMEWORK

The inhabitants of this country, the Sandawe, who, according to the 1957 census, numbered 20,000 in Usandawe and 28,300 overall, are one of the ethnographic puzzles of Africa. Usually, along with the Hadza living about 80 miles to the northwest, around Lake Eyasi,[18] they are classified as remnants of a Bushmanoid-Hottentot population that apparently was once fairly widespread over the eastern portion of Africa.[19] Several lines of evidence have been used to support this contention. First, both groups employ clicks in their languages, a fact that has led to their inclusion in the same linguistic family, Khoisan, as the click-speakers of southern Africa.[20] Second, the presence of some possible Capoid physical features has been alleged for the Hadza,[21] while the Sandawe are widely known for their distinct physical appearance, being on the whole shorter, lighter colored, and finer featured than neighboring peoples. Trevor made some morphological measurements among the Sandawe and concluded a definite Hottentot affinity.[22] Finally, there is the matter of hunting and gathering. The Hadza live almost entirely

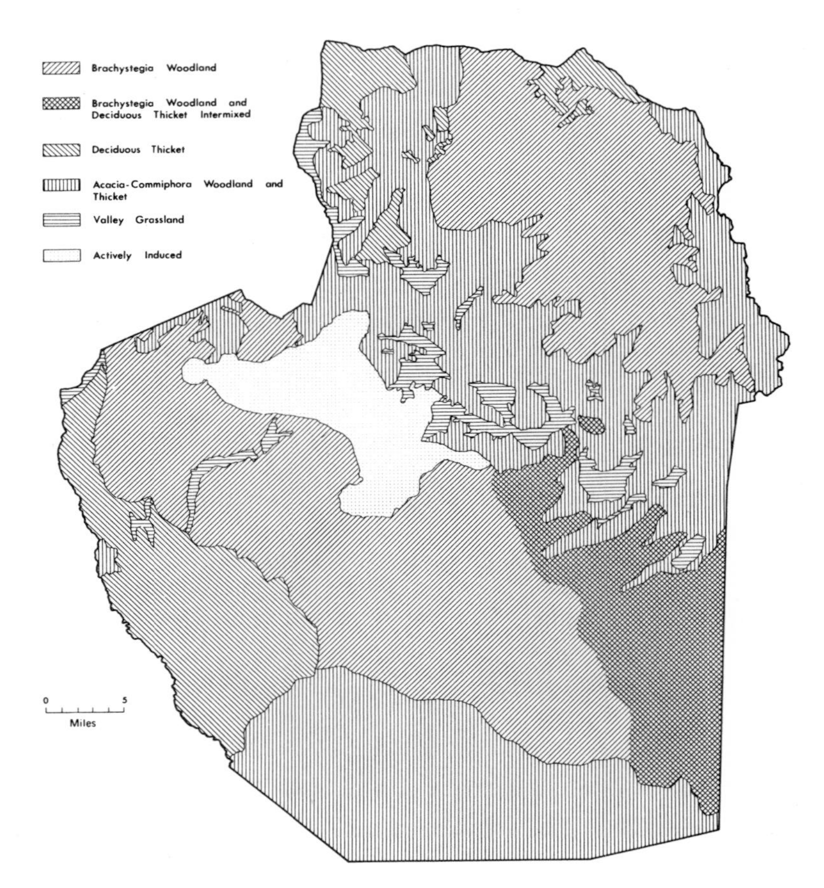

MAP 3 Vegetation assemblages of Usandawe.

off the products of the chase and the bush, engaging in little or no cultivation and possessing no livestock, not even chickens. The Sandawe, though subsisting primarily from agriculture and to a lesser extent from stock, have a strong hunting and gathering tradition. There is, in fact, much evidence to suggest that the Sandawe were essentially nomadic in the fairly recent past.

On the surface the evidence looks quite clear and straightforward. However, in line with the fate of most neat explanations in the social sciences, further inquiry suggests that some substantial revisions will have to be made. It has long been recognized that the linguistic link between the Hadza and

Sandawe is, at best, remote.[23] Now it appears as if there may not be one at all! A very recent analysis of pronominal and conjugational systems has shown that the Hadza language has virtually nothing in common with Sandawe.[24] In fact, it might not even belong to the Khosian group, but rather to the larger Afro-Asiatic family.[25] The suggestion would seem to be that clicks were merely incorporated, much as the Bantu Zulu and Bantu Xhosa of southern Africa have done, by way of contact with a truly click-speaking people, who, in this instance, unfortunately remain unknown.

The racial picture is equally in a state of turmoil. Outside of a few very gross features including steatopygia in women and shortness of stature, the Hadza appear to show no physical relationship to the Capoid group.[26] Cole even concludes that they are typically Negroid.[27] Furthermore, Trevor's work cannot be taken as the final word on the Sandawe. His sample from the population was quite limited and the comparative data presented exceptionally scanty. Also, he utilized only traditional anthropometric techniques. So far, no detailed analysis of blood groupings has been published. That the Sandawe are markedly different in some physical characteristics from other East Africans is fairly certain, but the direction and intensity of these differences have not been satisfactorily determined.

Thus, the evidence to support a Sandawe–Hadza connection and then ultimately to link the two peoples with the Bushmen and Hottentot has shrunk to negligible proportions. The Hadza, particularly, seem to stand apart, and though perhaps the Sandawe some day might be shown to have affinities with the southern African click-speakers, a considerable amount of research in physical anthropology, linguistics, and archeology, which as yet has not be pursued with reference to this topic, will be required before any more definite conclusions can be reached.

To complete our introduction, it will be useful to say a few words about the Sandawe's immediate neighbors. On the west, south, and east, are, respectively, the Bantu-speaking Turu (Rimi), Gogo, and Rangi (Map 4). The Turu (195,700, total population in 1957) are basically settled agriculturalists, though livestock enter as significant elements in the economic and, especially, social structuring of the tribe. The Gogo (population, 299,400) have a stronger focus on livestock than the Turu and are noted for the manner in which they have copied Masai customs and traditions. They verge on seminomadism during the dry season, when they must move their animals to the scattered patches of grass and water holes. Their agriculture is basically of a shifting variety. In contrast to the above two groups are the Rangi (110,300 in number). Livestock are not nearly as significant to the social order, and crop husbandry overwhelm-

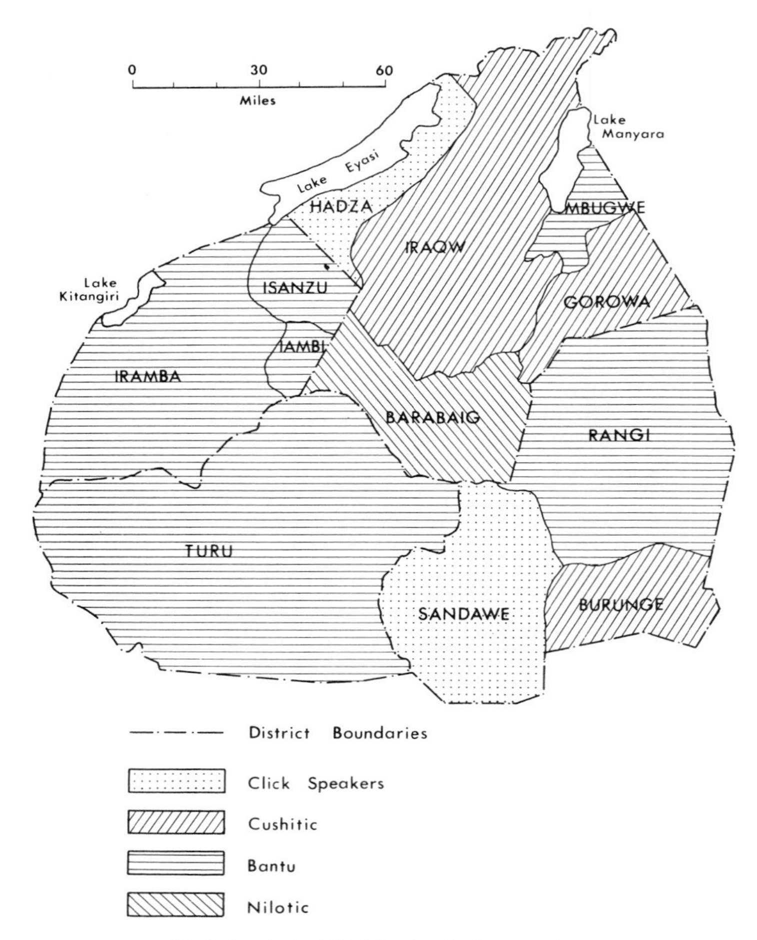

MAP 4 Central Highlands ethnic groupings.

ingly dominates their economic activity. In fact, the Rangi are usually classed as the best and most eager cultivators in central Tanzania. In addition, while both the Gogo and Turu have retained much of their traditional way of life, resisting a great deal of recently introduced technology and ideology, the Rangi have eagerly grasped at change. This is true with respect to religion, education, subsistence, and other areas. They seem anxious to equate themselves with anything new or modern.

Also to the east, just south of the Rangi, are the Burunge (Burungi). Not very much is known about this relatively small group (about 12,400 in total population) except their linguistic affiliation with the Iraqw (Mbulu), Gorowa (Fiomi), and Wasi farther to the north in the Central Highlands, and with the Mbugu of the Pare Mountains. Greenberg has classified the languages of these groups as Southern Cushitic, thus connecting them historically with Ethiopia and the Horn.[28] Whether or not this is correct is still open to much debate, but there appears to be little question that the five languages are closely related and form a unit when contrasted to other East African tongues. The Burunge themselves are shy and retiring, having little contact with the outside world. They have a heavy focus on livestock in their economic and social structures and crop cultivation is not highly elaborated, being similar to that practiced by the Gogo.

North of the Sandawe are the pastoral–nomadic Barabaig (total population, 30,600), a section of the more widely spread Tatog (Mangatti, Taturu). They are of Nandi stock, which makes them Nilo-Hamites or Southern Nilotes depending on terminological preferences.[29] Once the scourge of central and northern Tanzania as they wandered with their large herds of cattle and goats and raided for more, they have in recent years been subjected to ever-increasing pressures of settlement and absorption by the expansive-minded Iraqw.

Scattered in and around Usandawe are communities of Baraguyu Masia (Kwavi, Lumbwa), Nyamwezi, and Kimbu. The Baraguyu are distinguished from the main body of the Masai by their having adopted some crop cultivation.[30] Because of this practice, the truly pastoral Masai of the Masai steppe hesitate to admit that the Baraguyu are kin. Most of the Baraguyu in Usandawe, which seems to be their most northwesterly outpost in Tanzania,[31] inhabit the central lowlands belt east of Kwa Mtoro and the plains of the south. They too are termed either Nilo-Hamites or Southern Nilotes. The Nyamwezi and Kimbu are both western Tanzanian Bantu. They are predominantly cultivators, keeping little in the way of livestock beyond a few sheep and goats. The Nyamwezi have been the colonizers *par excellence* of Tanzania; their communities are found throughout the length and breadth of the country, primarily as a result of a heavy involvement in the slave and ivory trades of the nineteenth century. It was mainly ivory that seems to have brought them originally to Usandawe, and when the Germans arrived they found a thriving settlement at Kwa Mtoro plus several lesser ones.[32] Today the majority of Nyamwezi are localized in two places: Kwa Mtoro and along the road near Maxorongo. A few others are sprinkled about the countryside. The Kimbu first came as refugees from the slave trade, fleeing here, according to tradition, during the middle and latter portions of the past century. They are highly concentrated at Iseke, just outside Kwa Mtoro.

Lastly, there are the Arabs, also drawn in the nineteenth century by the slave and ivory trades. They are now almost exclusively engaged in shop-keeping, the majority of them living clustered together in the trading centers of Kwa Mtoro, Sanzawa, and Ovada. A few isolated Arabs are also found in the "bush" at Farkwa and Takwa.

II

CHANGING SANDAWE SUBSISTENCE PATTERNS

CHAPTER 3

THE HUNTING AND GATHERING PAST

During the latter part of the nineteenth century, when the Germans first penetrated what is now central Tanzania, evidence started to appear for the existence of a people—the Sandawe—who had recently begun a transition from a nomadic hunting and gathering way of life to one of settled crop and animal husbandry. The first to make the observation was O. Baumann, a leader of one of the German anti-slave missions of 1891–93.[1] The next was F. von Luschan, who undertook a brief ethnographic survey of the Sandawe as part of the remarkable scientific expedition of 1896–97 led by C. Waldemar Werther.[2] The same theme was repeated and elaborated in later works by Reche, Obst, Dempwolff, Bagshawe, and van de Kimmanade,[3] all of which led up to Murdock's classification of the Sandawe as East African Hunters.[4] In order to establish a base line for change, we shall have to examine the data behind this contention.

ORAL TRADITION AND LITERATURE

The Sandawe, themselves, tell of a time when they subsisted entirely off the products of the bush and chase; of how they traded meat, hides, and honey to surrounding peoples for grain and iron. They tell further of how, during their wanderings after game, they used caves and rock shelters for dwelling places in the rains, while in the dry season they constructed temporary shelters in sand watercourses and up in trees. This is all part of common oral tradition as handed down from generation to generation, and it would be difficult to find even a very young boy who could not repeat it. The Sandawe are convinced of their hunting and gathering past.

The same theme is also picked up among their neighbors. Turu informants cited how their ancestors found the Sandawe living "in the bush" without crops or livestock while many Gogo can relate the old trade that existed and speak of the Sandawe as people who received cattle "only yesterday."[5] An interesting story heard among the Barabaig is how when the Sandawe were first encountered it was necessary to beat them off because they looked on cattle as simply another type of wild animal to be hunted.[6] The notion of domestic stock had not yet been incorporated into their world view.

In Sandawe oral literature, which exists mainly in song form, the references to both hunting and gathering are marked by their great frequency, especially when contrasted to those mentioning facets of crop cultivation and stock-keeping.[7] The following is an example of a song, using a bird of prey—the kite—as an analogy, expressing the wish for a hunt to be successful.

Fly out, hey! fly out yes, fly out then.
Fly out hey! eh fly out, fly out.
Kite
(Chorus) What?
Oh kite!
(Chorus) What?
Fly out, etc.

Praise songs for individuals are widespread among the Sandawe, and nothing carries greater status in the society than to be praised as a skilled hunter, particularly an elephant hunter.

Oh Ziwa, he is a great hunter, father.
Ziwa indeed is a great hunter, well then,
(Hunting) party, party, father,
Ziwa is a great hunter, well then,
Ziwa is indeed a great hunter, father
Ziwa, Ziwa is a great hunter, well then,
(Hunting) party, party, father,
Oh Ziwa, he is a great hunter, father,
(Hunting) party, party, father,

Ziwa indeed is a great hunter, elder brother
Oh elephant, elephant, father . . .

Many more songs similar to the above could be cited, and though it is not possible to construct a total past culture from the evidence, still we can, I believe, get important clues as to the relative importance of various activities.

THE ROLE OF HUNTING AND GATHERING

The enthusiasm shown by the Sandawe for hunting and gathering, especially honey gathering, and the importance to subsistence of the products obtained from these activities have been commented upon by all the observers of Sandawe life. At the time of his visit, Baumann ranked bush products and game meat on a par with agriculture in providing food staples,[8] while 30 years later Bagshawe noted that "they are quite at home in the bush, upon the products of which they grow fat while other tribes are starving."[9] Van de Kimmanade's report of 1936 mentions that as late as the 1920's it was still possible to find some Sandawe living almost exclusively off game, bush fruits, and honey.[10]

Today, the Sandawe subsist primarily from domestic crops supplemented by livestock products, but hunting and gathering are still important activities and make substantial contributions to the food supply. Most men go on hunts at least several times during the year and many do so much more frequently. In fact, there are some individuals who spend a good share of their time in the pursuit of game, leaving the labors of the fields and the care of whatever domestic stock they might own in the hands of women and children. In a similar vein, honey collecting is a frequent and favorite men's occupation, while everyone from oldsters of both sexes through young children gather the various bush fruits and vegetables whenever they are available. But to establish the real significance of hunting and gathering, it will be necessary to look at the various activities in some detail and to make comparisons with the surrounding peoples when possible.

Three basic methods of hunting are used by the Sandawe: nets and spears, bows with poisoned or unpoisoned arrows, and traps of several kinds. The nets, termed *zágo,* range from 12 to 18 ft in length and stand about 3 ft high (Figure 9). They are woven from the fibrous bark of the baobab tree by men specially skilled in the art, though by no means do these individuals form what might be considered an occupational specialization or caste. Like all Sandawe craftsmen they work part-time, devoting most of their labor to food-producing and other mundane jobs. They learn their skill because of individual initiative and effort. There are no hereditary occupations. Net hunting is usually carried out during the rains, or shortly thereafter, when the game animals are most abundant and in their peak physical condition. An individual who possesses nets—they are costly and not everyone can afford them—and wants to organize a hunt goes to his friends and relatives to tell of his plans. If a big hunt is contemplated, then nine or ten nets will be needed, requiring the resources of several men. At the appointed time and place a party of 15 to 25, including women and children, will gather and fasten the nets to small-boled trees or prepared stakes. They then disperse to drive the game toward the barricade; beating with sticks, banging on pots and pans, shouting, and sometimes burning the grass. Several men are left behind to tend the nets, so that when the animals become entangled they can be speared, removed, and the nets set up again. Duiker, dik-dik, klipspringer, and bush pig are the primary victims. Larger-size animals are too strong for the nets to hold. When all is finished, the take is divided more or less evenly, except that the owner or owners of the nets are allowed to choose their favorite parts first.

Bow and arrow hunting is a much more individualized affair, especially during the dry season when the game is widely dispersed. The usual situation is for men to work in pairs or threes on a systematic hunt, though larger parties are organized for giraffe, buffalo, rhino, and elephant. Actually, we can probably consider Sandawe men as potential hunters most of their waking hours. A distinguishing cultural trait is that a man is sel-

FIGURE 9 A Sandawe hunting net (*zágo*) stands awaiting the drive of game animals.

dom seen without bow and arrow in hand, ready for a kill, should the opportunity present itself. Young boys are taught the art at age five or six, practicing on rodents and birds.

Several different arrow types are employed, depending on what is to be hunted.[11] The most common is the one termed *pándo* or sometimes *kasáma.* It has a laurel-leaf shaped blade and is used on all animals except the very large and the very small. For the latter, such as mongooses and porcupine, *supé* is constructed. It has a long, thin, tapering blade, usually of iron but of wood if for use by young children. The arrow to which poison is applied is known as *pagaré*. Ideally, poisoned arrows are preferred for any animal above the size of Grant's gazelle, for it is only in this way that a hit will ensure a fairly quick kill and thus save laborious and dangerous tracking. Traditional Sandawe poison is concocted from the saps of *Euphorbia nyikae* and *Excoecaria bussei,* but is rarely seen these days. Instead, a poison that is reputed to have greater strength is purchased from itinerant Kamba and Nguu traders. No comparative laboratory analysis has been made, so we cannot say in fact whether this is true. Conceivably, it could be just another example of the greater efficacy accorded to other people's magic. The blade on *pagaré* is triangular-shaped with the poison being placed about an inch below on the shaft. A very specialized arrow used only in hunting hyrax is *ála.* It is barbed, almost harpoon-like in appearance, and has a line attached, so that should the hyrax scurry into its hole after being hit, it can be lifted out. Finally, a blunt, wooden-headed arrow, *ísoli,* is for birds only. Metal-tipped arrows are much too valuable to risk losing on birds. Besides, with the exception of guinea fowl and francolin, most bird hunting is done by boys.

A variety of traps are also in use. Pits (*baán*) are dug, and sharpened stakes placed at the bottom, in the hope of impaling some animal. They are set around fields in an attempt to foil such marauders as bush pigs and baboons, or frequently they are situated along known game tracks in the bush. This is the primary method for hunting elephants, though it is necessary to use poisoned arrows to finish the kill. */Úmúku* is a snare-trap for small animals, mainly duiker, dik-dik, bat-eared fox, porcupine, white-tailed mongoose, and genet. A noose is placed on the ground and attached to a spring pole. When the animal steps inside the noose, a pre-set trigger mechanism is released, tightening the noose on one leg and suspending the victim in midair. For mongoose and genet the trap will be baited with a dead rat or mouse. A trap primarily for monkeys and guinea fowl is called *khíza* (Figure 10). It is simply a semicircular basket, some 45 in. long and

FIGURE 10 The trap *khíza* is used for monkey and guinea fowl. It is baited usually with maize kernels to lure the unsuspecting prey.

20 to 25 in. wide, made of any suitable thin branches and supported at the open end by a triggered pole. Maize kernels are placed inside as bait and when the monkey or guinea fowl enters, the pole is released, trapping him inside the basket. A similar device, *makíba,* operates on the same principle but substitutes a slab of rock for the basket. Rodents, guinea fowl, and francolin are hunted in this fashion. For small birds, an oval basket-trap, *thuundu,* is designed in such a manner that if a bird enters the small hole at the top to collect the bait of millet or sorghum, he cannot escape. There are no holds or sufficient space for the prisoner to maneuver. Young boys use a small net-trap called *kinóno,* also woven from baobab bark fiber, for hunting field mice and shrews. It is shaped similar to a cone, with an opening at one end which draws shut under the force of entry. Several of them will be placed along paths habitually traveled by the rodents so they can be chased in.

Appendix B gives a list of animals and birds hunted for food and the main methods employed.

In order to obtain at least a rough idea of how frequently wild meat is included in the diet, I asked one fourth grade student at each of six primary schools—Farkwa, Maxorongo, Kologa, Kurio, Sanzawa, and Ovada—to tabulate for three months everything served at mealtime in their homes. It was found that meat of all kinds was eaten at least once on 39 percent of days and of this total, 27 percent came from game. Unfortunately, it was not possible to record figures on quantity and thus total intake. We do know that at most meals at which meat is served, it is eaten in rather sparing portions, usually as a part of a relish with the ground grain, stiff porridge-like main dish known throughout East Africa as *ugali.* However, according to informants, when meat comes from game the size of dik-dik and up, it is common for the portions to be larger than average. In all probability, then, game meat is even more important quantitatively than qualitatively.

By comparison, the evidence from the neighboring tribes points to hunting as being a rather insignificant activity. For effective weapons, the Barabaig and Masai have only spears, and these are used on wild animals in most instances in protection for themselves and their livestock, or during certain ritual occasions when men have to demonstrate their strength and courage, or, less frequently, for mere sport. Their cultures contain no evidence suggesting a hunting past. About the same situation holds for the Burunge; they have neither bows and arrows nor nets and hunt only infrequently. Some trapping, mainly of birds, is engaged in, but on the whole, they take little interest in hunting, a fact mirrored by their general lack of knowledge about the habits of wild animals and the art of tracking. Today, hunting is virtually nonexistent among the Rangi. It may have been practiced in pre-European times on the greater scale, and its present negligible position might be due in large measure to the rapid expansion of population in this area during the past half-century and a consequent disappearance of most game animals and birds, yet there is little evidence in Rangi cultural traditions and patterns to suggest that hunting ever was of notable importance. Kesby reports that meat is not common to the diet, once a week being considered frequent, with varieties of game meat occurring very rarely.[12]

When we turn to the Turu and Gogo, somewhat more development in hunting is encountered, but both groups still fall short of the Sandawe pattern. The Turu employ the various traps and bows and arrows, though their proficiency with the latter evidently leaves much to be desired. According to von Sick, they cannot consistently hit anything smaller than a man-sized target at more than about 30 yards and consequently are no match for the Sandawe.[13] Furthermore, neither von Sick's early study nor the recent one by Schneider mention game as being at all an important supplement to the diet.[14] The Gogo are reportedly more competent bowmen and, in contrast to the Turu, nets are used. However, hunting nets seem to be heavily concentrated among the northern Gogo who have been in immediate contact with the Sandawe and, actually, there is a widespread tradition among them that suggests that the knowledge of how to use and make these nets was originally derived from the Sandawe.[15] Finally, among both the Turu and Gogo, hunting is an individual affair and is not considered a particularly prestigious activity. The majority of men, if they participate at all, do so infrequently.

It is true that some of the original Nyamwezi settlers came to Usandawe as elephant hunters, but at present they are overwhelmingly engaged in agriculture, and I found no evidence that they ever conceived of themselves as basically a hunting people. The Kimbu are even more attuned to cultivation and do not seem to do any hunting, at least in Usandawe.

Before closing the discussion of hunting, mention should be made of Sandawe dogs. They are used in tracking, chasing klipspringer up on exposed rocks where they make easy targets, and for killing monkeys and mongooses. Great pride is taken in possessing a good hunting dog and considerable boasting goes on between men as to who has the best. Songs are composed extolling their virtues and spoofing their shortcomings. On the whole, dogs are fairly well treated and cared for and their physical appearance makes a striking contrast to the scrawny, sickly looking ones kept by most East Africans. There is no evidence that any of the Sandawe's neighbors use dogs systematically for hunting.

With respect to gathering, the Sandawe are renowned in central Tanzania for their ability as honey collectors, and, as already indicated, a long-established trade has existed whereby the Sandawe have exchanged honey for grains and manufactured articles. Indeed, so enthusiastic are Sandawe men in their pursuit of honey, that they will immediately leave almost any other task as soon as the opportunity presents itself. Besides its value as an exchange commodity, honey is greatly desired as a food; it is eaten plain, used as a side dish with *ugali,* and as a sweetener and strengthener for their millet and sorghum-based alcoholic beverage. Termed *pombe* in Swahili, the Sandawe call it *k'amé* when it is brewed without honey, and *sakaláni* when honey is an ingredient.[16]

During the main swarming season of the honeybee, *Apis mellifera adansoni,* from March through mid-June, men are apt to go out every night either to their own hives or to help a friend or relative. In most instances, two or three hives will be tapped. If the hives are distant from the homestead–12 miles being about the maximum one-way distance traveled–the party is apt to remain away for three or four days at a time, collecting as much honey as possible. The owner of the hives always makes a payment of part of the produce or else of *pombe* to those who assist him. Some wealthy individuals, frequently possessing 100 hives or more, depend entirely on "hired" labor for collecting their honey.

The Sandawe watch their hives carefully, and when few bees are observed coming and going the combs are considered full and ready for tapping. The first procedure is to construct a fire from the branches of *Commiphora mollis* to produce a heavy smoke. A rope made from baobab fiber and weighted with stones at one end is then thrown up around a sturdy looking limb of the tree near the hive and one man climbs.[17] He is generally naked to prevent any bees from getting trapped between his body and clothing. The smoking branches and a basket called *tómbóto* follow. Next, the opening is located and the hive smoked for several minutes. A trial is made to see if the bees are groggy and back to one end of the hive. If the collector gets stung a few times, the smoke is reapplied until the bees' activity is reduced to a satisfactory level. When this stage is reached, the honeycombs are taken out one by one and placed in the basket, which is lowered to the party below. A good hive will yield about four gallons at a time during the main season and, hopefully, two tappings will be possible. To achieve this end, several honeycombs will be left behind to encourage the bees to remain and build again. In the dry season it takes three to four hives to produce the same amount and usually only one collection is attempted. When the men return to the household, the wax is removed and the honey stored in special gourds termed *nék'wa,* or, more recently, in kerosene tins. In former times the wax was discarded, there apparently being no traditional use for it. Now it is sold to local, usually Arab, merchants or to itinerant Indians.

Sandawe hives, *misiko,* are constructed from local soft woods, mainly *Albizzia tanganyicensis, Pterocarpus angloensis, Afzelia quanzensis, Commiphora ugogensis,* and *Commiphora caerulea.* A tree with a suitable bore is chosen (an average-size hive measures 50 to 55 in. in length and 18 to 20 in. in diameter), cut down, and the bark peeled off. The logs are split in half and hollowed out and fastened together again by twine. An opening for the collector's hand to enter is cut out, but is sealed off, except for a tiny hole allowing the bees access, by a removable panel known as *misiko sambala,* which appropriately translates as beehive door. After all these procedures are completed, the hive is strapped horizontally in the crotch of a tree, a feature that distinguishes the Sandawe from other Central Highland peoples who generally suspend their hives vertically (Figure 11). The important points to consider in finding a proper tree are that it should be tall, its bark should be smooth to prevent the honey badger (*Mellivora capensis*) from climbing, and preferably there should be no thorns to hinder the collector. Baobab trees prove the most suitable, while *Entandophragma bussei, Ostryoderris stuhlmanni,* and *Lonchocarpus eriocalyx* are also frequently utilized.

Wild honey is also sought after. Particularly desired because of its reputed sweetness is the honey obtained from the small melipona or stingless bees. The Sandawe recognize six varieties of bees—*ís'na, bakhóó,*

//axu, n!ukhee, k'amba tshina, and *ts'eme*–who build their hives in trees or shrubs and yield just a handful or two of honey at a time, plus two varieties–*torogombe* and *lok'to*–whose hives are found in the ground and produce yields of up to a gallon.[18] The hives are located most frequently by small deposits of wax left at the entrance, a wax the Sandawe use for sealing purposes, and sometimes the lesser honey guide (*Indicator minor*) is followed in its search for bee larvae. There does not appear to be the elaborate mythology attached to the bird that is found among the Bushmen,[19] but the Sandawe do say that one must be very careful in trailing it since its curiosity is aroused easily by anything unusual, including snakes.

Many other bush products are used for construction, medicine, and dietary supplements. With reference to food, baobab fruit is the most widely eaten, particularly by young children. Almost as popular are the fruits from *Salvadora persica, Vitex payos,* and *Berchemia discolor.* Special mention should be made of the leaves from *Gynandropsis gynandra, Ceratotheca sesamoides, Sesamum angustifolium, Amaranthus laburnifolia,* and *Corchorus trilocularis.* Together they form the most frequently used side dishes with *ugali,* being eaten at least once on 57 percent of the days, according to my findings, and constitute by far the primary source of vegetable greens in the Sandawe diet. A number of larvae, caterpillars, and termites are periodically consumed, while if severe crop failure occurs, reliance is placed upon several varieties of roots, seeds, and fruits that are not normally part of the food supply. A tabulation of the Sandawe's most common and important bush products is provided in Appendix C.

FIGURE 11 Three Sandawe beehives are cradled in this baobab. Note the tree's fruit, the seeds of which are eaten and the pulp used in brewing.

There is little in the way of a system to food gathering. As mentioned previously, everyone participates, though actually women and children do the greatest share. Much of it might best be called random collecting, since it is performed while in pursuit of other activities, such as herding livestock, picking up firewood, and going on visits. Materials for building and medicines are gathered as they are needed. These are jobs for men only.

The use of bush products is not really a particularly helpful index of a former hunting and gathering existence. Honey is widely desired and collected where conditions permit, though admittedly the enthusiasm and probably also the skill shown by the Sandawe are hard to match. Furthermore, most African people, including certainly the Sandawe's neighbors, seem to possess a highly developed plant lore and use a wide variety of the available vegetable resources.

What might be revealing, though, are the attitudes expressed about the bush. To the Sandawe, it is virtually synonymous with the best place to live; there should always be plenty of trees and shrubs near a man's home. Several old informants who had visited places outside Usandawe expressed amazement at how the Turu, Rangi, Gogo, and others could live when the landscape was a continuous succession of fields, grazing grounds, and homesteads. "Where would the firewood, building materials, honey, and fruits come from?" they asked. "Where would the animals live?" To them it appeared a most undesirable form of existence. A good cover of "wild" vegetation is considered necessary to make a place properly habitable. Such sentiment does not appear, even in rudimentary form, among the cultivators and stockkeepers surrounding the Sandawe, and, indeed over most of Africa, one is left with the impression that man and extensive stands of trees and shrubs are incompatible.

THE UNDERDEVELOPED HUSBANDRY TRADITION

When we take a close look at crop and animal husbandry practices, the Sandawe demonstrate a lack of elaboration and sophistication that might be expected of a people who have not in the past been highly dependent upon them for gaining their livelihood. Even by central Tanzania standards, which are none too high, they are inevitably placed last in their ability as farmers. Kondoa District Agricultural Reports constantly mention their inefficiency and the little effort they put into cultivation. Bagshawe stated that "the tribe are not skilled or energetic agriculturalists,"[20] while a report for the development of central Tanzania in the 1950's was even more emphatic, calling them just plain "poor" farmers.[21] Particularly revealing are statements by a number of Rangi and Gogo male informants, who said they would never consider taking Sandawe women for wives because "they do not know how to cultivate properly."

The timing of various agricultural operations seems especially haphazard and unsystematic. There is no structured routine as has been well-documented for many African peoples, including the neighboring cultivators. The clearing of new fields can begin as early as March or as late as October, while burning is seen any time from August up to the beginning of the rains (Figure 12). Ideally, in late November or early December, an attempt is made to dry plant the main fields, which consist of a mixture of bulrush

FIGURE 12 This field was newly cleared in June. It will be fired sometime from late October to early December.

FIGURE 13 A field in early December standing ready to be planted. The stubble in the foreground indicates this section was used the previous year, while the ash piles in the background mark a first-year plot.

millet (*Pennisetum typhoideum*), sorghum (*Sorghum vulgare*), maize (*Zea mays*), pumpkins (*Cucurbita moschata, Langenaria siceraria, Langenaria sphaerica,* and one undetermined), haricot bean (*Phaseolus vulgaris*), and recently castor (*Ricinus communis*) (Figures 13, 14, 15). In fact, even if the rains break on schedule, the initial planting frequently goes on through January. The timing is inconsistent even among particular individuals. Planting can be early one year and late the next, depending on when there is "sufficient time," as men frequently say. The same situation is encountered on the two occasions for weeding. The first is supposed to be performed when the green heads of the newly emergent crops are about two or three inches above the ground. This is a labor-intensive

FIGURE 14 The same field as in Figure 13, showing a man preparing the grain seed holes with a hoe.

FIGURE 15 After the field has been prepared, the women of the family plant the grain seeds by dropping a mixed assortment in the holes and kicking dirt over them.

operation in which neighbors work together on each other's fields (Figure 16). The price is *pombe,* or sometimes meat, at the end of the working day—usually about 6 hours in length. A single field, depending on its size, can take up to a week, there being no rush to get the job finished. Consequently, a time lag of up to two months often will ensue before the entire round of fields is taken care of. Quite understandably, the last ones on the list suffer from weed encroachment. The second weeding is not a communal affair, each family seeing to its own fields. Only two or three days of half-hearted effort are expended, usually sometime from late March to early April, about a month and a half before the harvest begins.

After the first weeding of the main fields, the preparation and planting of groundnuts (*Arachis hypogaea*), Bambarra nuts (*Voandzeis subterranea*), sweet potatoes (*Ipomoea batatas*), and cowpeas (*Vigna unguiculata*) is begun. Most homesteads attempt to grow at least three of the four in any one season, but there is great variability from family to family and from year to year. Frequently, the time of planting will have passed because of slowness in the preceding occupations; also, the expenditure of necessary effort is rather distasteful. The soil for groundnuts and Bambarra nuts must be completely turned before planting, while ridges have to be constructed for sweet potatoes and cowpeas (Figure 17).

Such a degree of seeming disorganization and actual apathy with regard to agricultural work is not found among the other cultivating peoples of central Tanzania. Especially among the Rangi and Turu, one encounters a much more definite phasing to the various activities and there is little question that during the times of field preparation, planting, and weeding

FIGURE 16 Neighborhood parties perform the first weeding of the grain fields. Each family is responsible for providing food and drink while the group works in its fields.

all other endeavors are subsidiary. Not having "sufficient time" would be unheard of except in circumstances such as severe illness.

Also, the Sandawe have not elaborated a system of land tenure. Land is divided up and named according to subclans, each subclan being oriented around a particular hill (*gáwa*) which bears its name and for which it claims an ancestor as "owner," but this division has little practical meaning. A hill usually has people from entirely different subclans living around it, and many Sandawe are unsure of where their hill is located.[22] People move freely and often frequently, and as long as there is land available it is open to anyone, including non-Sandawe. The concept of clan ownership, as practiced by the Turu, Gogo, Rangi, and many African groups, where formal permission to take up land must be obtained from clan elders who are responsible for a specified segment of territory, is unheard of.[23] The only rights are those vested in the individual once he takes up cultivation. He cannot be dispossessed of his land, as is possible under systems of clan ownership (though, of course, he can be run off if his neighbors dislike him), and his rights to a claim lapse only when it is no longer put to use. When the land reverts to bush, it can be reclaimed by anyone, there being no maintenance of fallow like that noted for the Turu and Gogo.[24] However, if a man plans to use a piece of land again within a few years, all that he has to do is let it be known and no one would likely claim it.

Another feature of the Sandawe's rudimentary system of land tenure is the lack of readily visible markers bounding adjacent fields. This generalization holds even for the Ovada and Kurio-Kwa Mtoro areas where the trend is toward permanent cultivation. The ditches, sisal hedges, and thorn brush

FIGURE 17 Sweet potato fields are usually prepared after the grain fields have had their initial weeding. The dogs are typical of those found among the Sandawe.

employed by the Rangi, Turu, and Gogo are rarely encountered. Sometimes paths are used as markers, but fields often merge into one another and only the respective owners know where one man's property leaves off and someone else's begins.

A fairly safe conclusion, then, is that the Sandawe have very little concern over rights in land for cultivation. As corollaries of the above, land, with a few exceptions to be noted later, is not inherited, is almost negligible as a source of dispute, is not a topic of much conversation, and there is no evidence to suggest a desire to accumulate it for wealth. We might go so far as to say that the Sandawe show about as much interest in the distribution and exploitation of agricultural land as the neighboring peoples do in hunting.

When animal husbandry practices are examined, further striking comparisons emerge. East Africa as a whole is generally known for the extent to which livestock, particularly cattle, enter into the economic and social lives of the people. Even excluding the pastoralists, a great share of activity is oriented toward acquiring as many animals as possible, for this leads the way to status, prestige, and power in the society. In central Tanzania, the Turu and Gogo are classic exemplifiers of the pattern and, since fairly good documentation exists for each, they can be used as a base against which to compare the Sandawe.

In both groups the possession of large numbers of livestock is vital to a man's economic and social standing; indeed it is the only way he can make his voice heard in the affairs of the group. Livestock are the symbol of wealth and property. As one Turu saying collected by Schneider puts it, "cattle are our banks, our stores, our farms, our wives, our clothing, everything."[25] Thus, the vast majority of homesteads have at least a few animals and a desire to add more and more. Using the 1965 estimates by the Ministry of Agriculture of livestock population, the Turu average about 1.4 head of cattle and 1.1 of sheep and goats per person and the Gogo 2.4 and 2.2, respectively. So strong has the obsession with numbers become, that slaughter rarely occurs. The Gogo have carried this to such an extreme that reportedly during times of severe famine they have even refrained from killing goats.[26] It might be said that it takes an act of God or the ancestors to pry an animal loose, for it is only on important ritual and ceremonial occasions that slaughter is regularly performed. In fact, it is probably the close attachment between livestock and the supernatural that, at least in part, accounts for the desire to accumulate large herds. The more resources a man has to propitiate and pacify the spirits the greater will be his power and standing in the community.

A tremendous lore about livestock has been built up. The Turu have phrases and songs of praise about cattle, extolling their strength and beauty. Men and bulls are compared, while women and cows are equated at various phases during the life cycle.[27] To the Gogo, behavioral mannerisms by livestock are taken as omens for the future,[28] while they have built up an elaborate vocabulary for designating cattle on the basis of age, size, color, horns, and ears. An average herdsman will know 40 or 50 such terms and each animal will be separately designated according to some combination. Similarly, both groups recognize a variety of livestock diseases. For instance, the Gogo have at least 20 descriptive terms and specific herbal medicine to treat each disease.[29]

Grazing is a highly regulated activity. The Turu have partitioned off the grassland mbugas, each man having the right to use so much. The areas are demarcated and fenced as thoroughly as cultivated land and there is a considerable amount of dispute and litigation about who has what rights where.[30] Among the Gogo, though most uncultivated land is considered free range, small, especially favored spots are often enclosed with thorn hedges to provide grazing for very young and very old animals.[31] These enclosures are treated as private property. Also, during the dry season the Gogo assume a seminomadic existence. The men move off with their herds to search for pasture and water, setting up temporary camps along the way, while the women remain behind and care for the original homestead.

To the Sandawe, much of the above is alien. First off, the numbers of livestock and proportion of owners are fewer. If you consider solely the total population resident in Usandawe as the base, this fact will not be obvious, since figures of 2.7 cattle and 2.1 small stock per person are attained. However, a good portion of the livestock in Usandawe is actually in the hands of non-Sandawe. In the Farkwa and Sanzawa subchiefdoms well over half the animals are owned by Masai and Gogo, while in the other two subchiefdoms—Lalta and Poro—many are owned by Turu. An optimistic estimate would cite one head of cattle and one sheep or goat per person, with between 40 percent and 50 percent of the homesteads owning no domestic animals other than chickens and dogs.

Even more revealing for our purposes are the attitudes and practices surrounding livestock. Livestock are a facet of wealth for the Sandawe, yet by no means are they all-dominant, and they certainly are not the primary arbiters of a man's position and prestige in society. Many locally important people are virtually without stock and, interestingly, a widely expressed opinion is that stock are not worth all the trouble involved in keeping them. When Sandawe were questioned about what makes a man a leader, the almost universal reply was that he demonstrate success in his choice of concentration, be it hunting, honey collecting, farming, herding, or whatever, and that his advice prove consistently useful.

Most slaughtering, as with the Turu and Gogo, is done in connection with rituals and ceremonies, but there is no real compunction about killing for food at other times. This is especially true for goats.

In addition, Sandawe livestock lore and mythology are rather poorly developed. Perhaps some 30 to 35 terms relating to cattle can be found in use throughout Usandawe, but the average herder seems to know only about five or six. The most I found recited by a single individual was 12. Terms in common parlance are listed in Table 1. Only three disease terms are widely used: *k'wana* is for sickness in general; *sakílo* refers to anthrax; and *múkuzi* apparently designates mange. The one potion that was found in use is made from the sap of *Commiphora swynnertonni.* Praise songs involving livestock are comparatively few and the equation between men and bulls and women and cows does not seem to be made. Sandawe con-

TABLE 1 Common Sandawe Cattle Terms

Description	Term	Description	Term
Cattle in general	*múmbu, húmbu*	Spotted equally white and black	*xolá*
Bull	*k'ámba*	White	*phoo, solí*
Ox	*áfure*	Light gray	*!ukli*
Calf	*dáma*	Dark gray	*zanga*
Barren cow	*ts'aatha*	Reddish brown	*duxahní, mori*
Solid black	*k'ánk'ara, titi*	Reddish brown with black stripes	*samú*
Black with white face	*khiréka*	Reddish brown with white patches	*baraxáti*
Black with small white spots	*bás', légendi*	Tawny	*butl', árara*
Black with larger white spots	*/ili*	Tawny with black spotting	*xoporu*
Black with large white patch on side	*nís'*	Tawny with white patches	*omí*
Black with a white blaze	*mará*	Horns bent down	*koróngo*
Black with white patch on back	*zinzál*	Horns bent forward	*xóngo*
Black with white patch on throat	*légini*	Exceptionally small horns or hornless	*!hwegé*

ceptions of beauty and strength are more likely to be expressed with game animals as the literary metaphor. The following two songs provide a general example:

> The giraffe, oh alas, the gracefully-patterned animal in the land of Kirange, she is lying silently.
> Oh the giraffe, oh alas, the animal lies there still,
> Oh woe, in the endless plain-country she, the beautiful one lies there silently.
> Oh giraffe, oh alas, the animal lies there still, mother, in the endless plain-country she, the beautiful one lies there silently.
> The giraffe, oh alas, the gracefully-patterned animal. Oh woe, in the endless-plain country she . . . [and so on]

> Yes Mudé indeed, she is like a klipspringer, that child has a graceful swagger, Mudé!
> Mudé indeed, she is like a klipspringer.

Special grazing enclosures and competition for grazing land are unknown. All land, except, of course, cultivated fields, is open to grazing by anyone who so desires. The animals are taken to and from the homestead each day, even during the height of the dry season. Unlike the Turu and Gogo, grass burning to produce young, green rhizomes is not performed. On the whole, then, the Sandawe appear to have elaborated the technology of livestock-keeping about as little as that of crop cultivation. There is a definite under-developed husbandry tradition here.

OTHER INDICATORS

The Sandawe house type is the typical rectangular *tembe* found throughout central Tanzania (Figure 18). Its general features have been described on many occasions and consequently I shall not bother to elaborate on them again.[32] However, another dwelling unit exists, one which appears by all indications to be more traditionally Sandawe. Termed *khambi* or *kirĩngóó,* it is employed now as a temporary shelter by hunters who go out on expeditions lasting more than several days. Tradition implies that at one time it was the only constructed shelter, and Obst observed during his travels that it was still used by some families as their main house.[33] A *khambi* is constructed by piling dried grass and bark over pliable branches which are bent into a conical frame. It greatly resembles the hut presently used by the Hadza, and to my knowledge nothing similar can be found among the other nearby peoples.

FIGURE 18 The overwhelmingly dominant house-type in Usandawe is the *tembe*.

The Sandawe demonstrate a catholicity of diet which one would expect of hunters. As Bagshawe has put it, "they will eat just about anything,"[34] including lions, leopards, cheetahs, serval cats, civet, mongooses, shrews, baboons, monkeys, snakes, jackals, and even, as one informant admitted in an evident moment of weakness, hyenas, if they are hungry enough. Lions will be chased off kills, and carrion will be eaten if it can be determined that the animal died of some natural causes like snake bite. This is strikingly opposed to the practices of the neighboring peoples, who often speak derogatorily about the Sandawe's food habits and who, in fact, have some very specific food avoidances. Masai warriors are forbidden the meat of wild animals in total, while the Burunge, Rangi, Gogo, and Turu restrict their preference to hooved beasts and birds. Those with claws or nails, such as the cats and primates, are taboo. The Turu traditionally were not allowed to partake of the flesh of game animals in the house. It had to be cooked and eaten out-of-doors.[35] Another common avoidance is fish, the rationalization being that they resemble snakes, which are feared not only for their venom but also for their association with sorcerers. The Sandawe are avid fishermen, and really were the only fishermen of any note in central Tanzania until recent programs designed to overcome the reluctance to fish as food were instigated.

Though systematic analysis of the Sandawe language is just beginning, some extremely interesting preliminary results bearing on the matter of a hunting and gathering past have been revealed. It seems that with but one exception, all words relating to the various crops grown are of non-Sandawe origin (Table 2).[36] The exception is one of the three terms for maize—*n/íni*—but this has the wider meaning of short or stunted in stature, which maize certainly is when contrasted to the sorghums and bulrush millet. Further, Dempwolff has shown that much of the vocabulary referring to food preparations, agricultural implements, and agricultural activities are of Bantu derivation.[37] In the same vein, the livestock terms presented in Table 1 are basically non-Sandawe.[38] Only the four traditional color designations *k'ank'ara, phoo, butl',* and *zã̃nga* are excepted. As far as can be determined, all wild animal and plant terms are distinctly Sandawe.

Finally, reference should be made to ritual associations. It is generally held in anthropological circles that ritual mirrors those elements of the

TABLE 2 Crops Grown by the Sandawe

Common Name	Scientific Name	Sandawe Name
Bulrush millet (two varieties)	*Pennisetum typhoideum*	*!ekóo* (spineless) *sóx'mo* (spined)
Sorghum (four varieties)	*Sorghum vulgare*	*gorombáa* (red) *tonóntoo* (red) *phémba* (red) *lalángaa* (white)
Maize	*Zea mays*	*n/íni, áana, mambatú'a*
Finger millet	*Eleusine coracana*	*beréẽ*
Groundnut	*Arachis hypogaea*	*kalangá*
Bambarra nut	*Voandzeia subterranea*	*kozigá*
Haricot bean	*Phaseolus vulgaris*	*malaage*
Cowpea	*Vigna unguiculata*	*kósa*
Sweet potato	*Ipomoea batatas*	*mphókaa*
Cassava	*Manihot esculenta*	*moógo*
Pumpkins and melons	*Cucurbita moschata* *Lagenaria sphaerica* *Lagenaria siceraria* ?	*amphaní* */'úr'taa* *koóngo* *tangá*
Rice	*Oryza sativa*	*mchele*
Sugar cane	*Saccharum officinarum*	*magúa*
Tobacco	*Nicotiana tabacum*	*tumató*
Castor	*Ricinus communis*	*sa'ú* (spined) *thándiko* (spineless)

culture which are, or traditionally have been, of primary importance.[39] The more intense the ritual surrounding a certain event the more likely it is that the event is closer to the core of the culture. I cannot speak with much authority on Sandawe rituals, but from the testimony of informants it seems fairly certain that there are many more centering around and invoking aspects of hunting and gathering than crop and animal husbandry. Circumcision, for example, is undoubtedly the single most important ceremonial occasion for the Sandawe. It is the time when boys and girls cease to be merely children and become fully initiated members of the society, with all the attendant rights and responsibilities. Besides merrymaking and feasting, there is much serious educational instruction in important tribal traditions. For this and for the actual act of circumcision, the eligible boys and girls, usually about ten of each, retire to separate camps. In the boys' camps, which is all I have information about, stress is put on learning two things, songs and hunting. The boys are taught to make bows, arrows, spears, and traps and how to be proficient in their use. The habits of the various animals and the art of tracking are gone over, and the whole lore of living in the bush is recited. Virtually no attention is paid to cultivation or herding. By way of comparison, there is evidently considerable stress on cattle in the male circumcision ceremonies of the Turu and Gogo. For example, when a Turu boy finishes his ordeal he is given a heifer by his father.[40]

Various medicines, potions, and offerings to ancestors and respected elders are employed to ensure success in hunting and honey collecting. Again, similar ceremonies for crops and domestic animals are negligible. Nothing resembling breaking of the earth, first fruit, and other common agricultural ceremonies was detected. Rainmaking is practiced, and it certainly is an important ritual occasion, but by no means can this be considered an exclusive possession of long-settled cultivators or herders; rain is equally as important for game and bush products.

Thus, the evidence from oral tradition, historical documentation, and from present subsistence patterns, technology, dietary habits, cultural attitudes, linguistic usages, and ritual associations seems, when taken together, to support the contention that the Sandawe once followed a basically hunting and gathering way of life. Though it is not possible to close the case with absolute certainty, even the skeptic would, I think, admit that at the very least wild game, honey, and other bush products once played a very dominant role in overall subsistence. What now must be investigated is how and why crops and livestock came to supplant them.

CHAPTER 4

THE ADOPTION OF CROP AND ANIMAL HUSBANDRY

According to Hogbin, the influential events that tend to produce social and cultural change include the following:

1. Dissatisfaction with the existing social order
2. Alterations in the density of population
3. Modification of the natural environment
4. Migration to a new territory, and
5. Invasion by foreigners possessing superior tools[1]

Murdock has compiled a basically similar list, which reads as follows:

1. Increases or decreases in population
2. Changes in the geographical environment
3. Migration into new environments
4. Contacts with peoples of differing culture
5. Natural or social catastrophe—floods, crop failures, wars, epidemics, and economic depressions
6. Accidental discoveries, and
7. Biographical events as the death or rise of a strong political leader[2]

Of these various conditions, contact with foreigners, or, if you prefer, differing cultures, is generally conceded to be by far the most significant.

According to one study, "civilization is largely the accumulative product and residue of this ever widening process of culture contact, interchange and fusion,"[3] while Childe has stressed that the most important element in man's environment is his fellowman, since culture basically grows through the process of borrowing from others.[4] It will be demonstrated how the adoption of crop and animal husbandry by the Sandawe apparently is another example of this process in action, with population growth and food shortages as important related occurrences.

Hunters and gatherers require isolation in order to maintain their identity. "They survive . . . only because temporary sanctuary remains to them in lands so desolate or difficult that neither cultivators nor herdsmen, nor the peoples of advanced technologies have found use for them."[5] Competiton means destruction, either by the way of outright physical annihilation or assimilation. Consequently, a Kalahari desert, a tropical rainforest, a central Australia, a special niche in a larger society, or some other area of retreat is necessary to keep them going. In Usandawe there is nothing to hide behind. The Itigi thicket could possibly qualify as a barrier, but it only affects the southwest and can be skirted easily. The Sandawe Hills are not nearly rugged enough for effective isolation, though in conjunction with the scatter of population they have probably helped to hinder the rapid spread of innovation. For all intents and purposes, then, Usandawe can be considered as relatively open to movement and communication. All that is needed is an active source and a stimulus.

The available evidence suggests that the Turu provided the source for the initial introduction of domesticated crops to the Sandawe. A widespread Turu claim is that "our ancestors showed the Sandawe how to cultivate," while it is a common tradition among the Sandawe that it was from the Turu that they first learned about growing crops. This notion is tied in with a Sandawe story of how, during a time of severe famine, both peoples wandered in search of food, some Sandawe eventually settling among the Turu, some Turu settling in Usandawe. Thus, the Sandawe were able to see for the first time how crops could be grown. In all probability, the tale is an oversimplification, but it does seem to contain at least an element of truth, for the Sandawe and Turu have intermixed to a considerable degree; a fact noted even by the earliest observers. Baumann pointed out how Turu frequently migrated into Usandawe and concluded that they were really the only outsiders with whom the Sandawe had much intercourse,[6] and somewhat later van de Kimmanade remarked on the extensiveness of the intergration of the two peoples in the Kurio-Ovada area.[7]

Map 5 portrays the present-day, overall distribution of population in

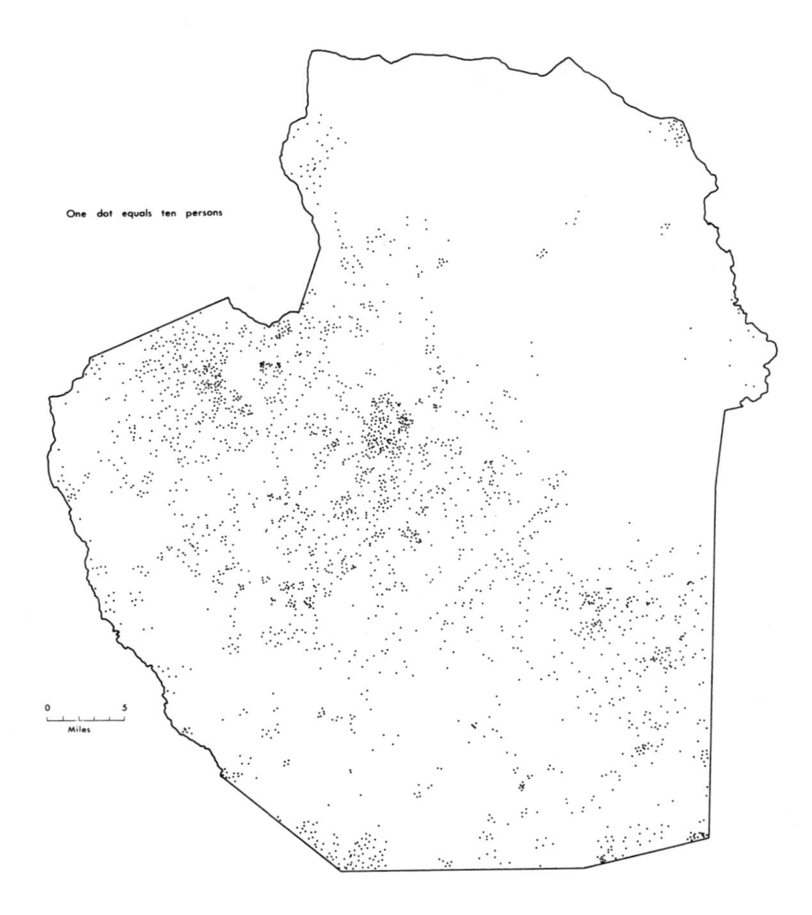

MAP 5 Population of Usandawe.

Usandawe, while Map 6 and Table 3 break it down by non-Sandawe components. The data essentially confirm the above observations. Turu are heavily represented, and, in fact, they dominate some portions of Usandawe, particularly the northwest. There are many Sandawe place names and some subclan hills in this area, a situation continuing westward into Singida district below the Mponde escarpment and indicating that all of it was probably once inhabited by Sandawe. Today few can be found, and the northwest can best be considered as effectively part of Turu country.

In conjunction with proximity, there has been a fair amount of fusion and intermarriage. Several Sandawe clans claim Turu descent, while one or two Turu clans ascribe their ancestry to the Sandawe.[8] Table 4 summarizes the results of a survey I conducted, where each child in the third and fourth grades of the primary schools in Usandawe was instructed to list the ethnic affinities of his parents and grandparents. It is immediately apparent that

TABLE 3 Percent Composition of Non-Sandawe in Usandawe

Tribe	Percent of Usandawe Population	Percent of Non-Sandawe Population
Turu	14	50
Gogo	4	18
Baraguyu	2	10
Nyamwezi	1	6
Kimbu	1	5
Rangi	0.5	3
Burunge	0.2	1

Source: East African Statistical Department, *General African Census, August 1957, Tribal Analysis, Part II, Territorial Census Areas* (Dar es Salaam: Government Printer, 1958), pp. 94–95.

TABLE 4 Rates of Intermarriage between Sandawe and Non-Sandawe from Parents and Grandparents of Primary School Children

Parents		Grandparents	
Combination[a]	Percent	Combination[a]	Percent
Turu–Sandawe	40	Turu–Sandawe	43
Sandawe–Turu	28	Sandawe–Turu	33
Kimbu–Sandawe	9	Rangi–Sandawe	7
Gogo–Sandawe	9	Gogo–Sandawe	5
Sandawe–Rangi	4	Sandawe–Rangi	2
Sandawe–Kimbu	2	Sandawe–Gogo	2
Sandawe–Gogo	2	Sandawe–Kimbu	2
Rangi–Sandawe	2	Sandawe–Burunge	2
Nyamwezi–Sandawe	2	Barabaig–Sandawe	2
Barabaig–Sandawe	2	Nyakyusa–Sandawe	2

[a]Man's affiliation listed first.

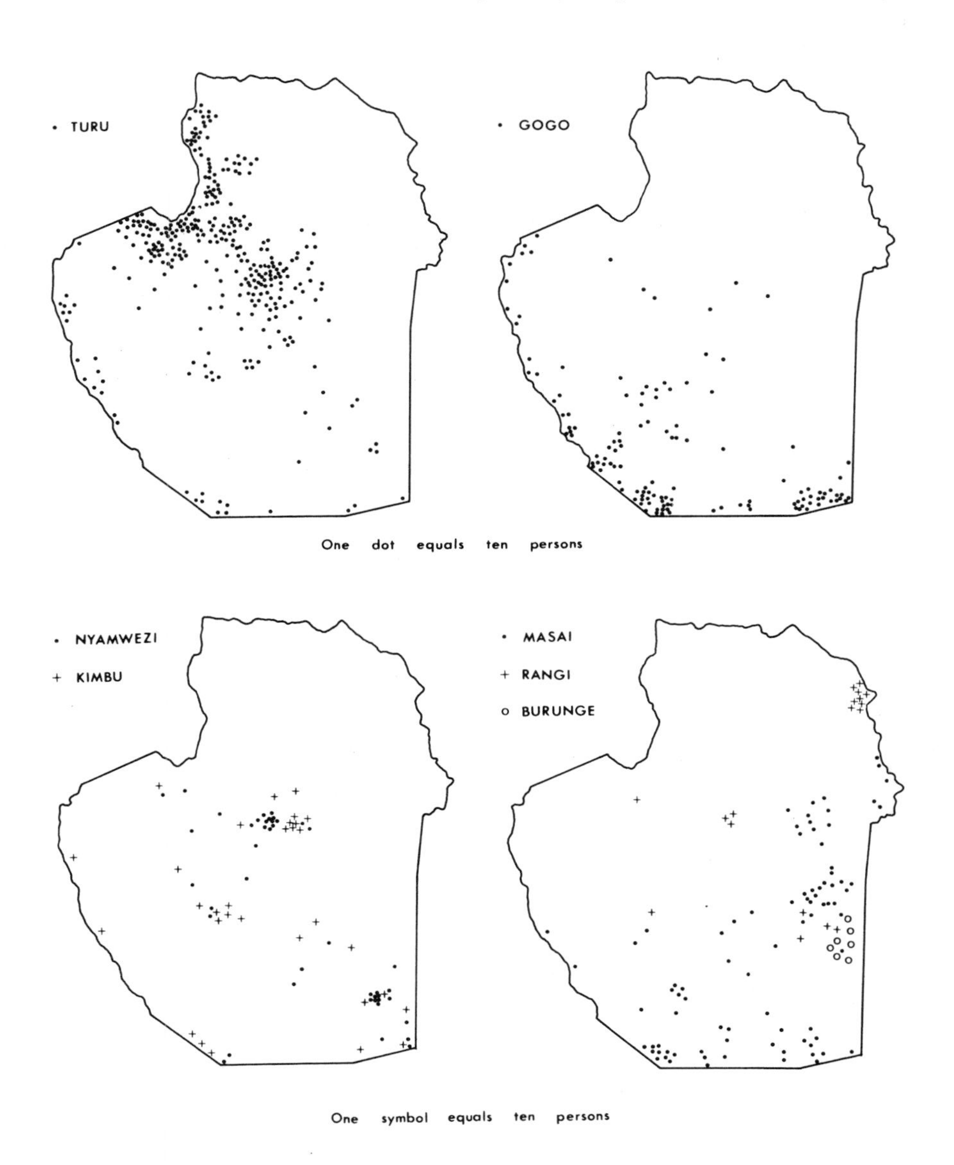

MAP 6 Distribution of non-Sandawe in Usandawe.

the overwhelming majority of marriages of Sandawe with non-Sandawe occurred with the Turu. Social barriers to intermarriage between the two groups are apparently lacking, and, according to informants, the process has been going on for a longer time than two generations. As one old Turu put it, "There has never been any hostility between us; we are as one people." Contiguity and degree of interrelationship seem to support the contention of a Turu influence on the Sandawe.

Another facet indicating a Turu influence is that Sandawe crop cultivation reaches its maximum elaboration where contact with the Turu has been most intense. Permanent fields, manuring, and more careful attention to the cycle of agricultural activities are characteristic in the Kurio-Ovada area. Several informants revealed they were cultivating the same plots that their fathers had used. Land is even becoming important enough to be heritable.

Constituting a possible secondary source for the spread of crop cultivation to the Sandawe are the Gogo. The portion of Usandawe southeast of the Bubu River, where contact with the Gogo is now most extensive (as it undoubtedly was also in the past), has always been peripheral to Sandawe settlement. Movement in significant numbers in this direction seems to date only from sometime during the middle of the last century, when tradition has it that another severe famine forced the people to migrate in search of food. However, one segment of the Sandawe, known as the *Bisa,* apparently had occupied the southeast for a much longer time. They are reputed to have been primarily hunters and gatherers, with only a minimal amount of cultivation and no livestock, when the trans-Bubu movement took place—a time when the bulk of the Sandawe were seemingly well on their way to a more settled life. Reche commented on the comparatively primitive nature of these "southern" Sandawe,[9] and today one still finds the technology of agriculture least well developed here. But getting back to the initial introduction of crops to the Bisa, the Gogo would appear as a conceivable choice. There has been some intermarriage between the two groups, and also there are a few clans on both sides who claim derivation from the other group. The Sandawe have no objections to mixing with the Gogo, except for a general fear of their powers of witchcraft. They frequently term the Gogo rather "dangerous people."

In any event, no matter whether from the Turu or from the Gogo, cultivation would have centered around bulrush millet, sorghum, Bambarra nuts, cowpeas, and the various pumpkins and melons. These are the long-standing, traditional crops throughout central Tanzania, others being either of minor importance or of recent introduction. The bringing of maize is attributed in Sandawe tradition to the Nyamwezi and seems to fit in with the fact that of all the people found in and around Usandawe, only they have grown it for a long time as a staple. It is also likely that the Nyamwezi brought sweet potatoes and haricot beans. Both crops require ridging and the Sandawe are quite uniform in claiming that it was the Nyamwezi from whom they learned ridge construction. Interestingly enough, the Nyamwezi ridge other crops as well, including maize. Rice, a very minor crop, is linked with the Arabs. The main evidence comes from oral testimony and is supported by the exclusive use of the Swahili word *mchele* for rice. Tobacco seems to be of fairly long-standing cultivation, for even in von Luschan's time it was widespread.[10] Nothing has been uncovered regarding its origin, however. The same situation holds for eleusine. It is an old African cultivated crop, but its original source of introduction into Usandawe remains a mystery. Neither groundnuts nor sisal were mentioned as being grown by observers from Baumann (1891) through Bagshawe (1920's). The presence of both crops is a result of government policy under the British administration and will be discussed at greater length in the next chapter.

Employing linguistic evidence and clan histories, Tenraa is in the process of attempting to trace the entrance of livestock into Usandawe.[11] It has already been pointed out that the terminology in use is not of Sandawe derivation. Now, it seems fairly certain that we can ascribe most of these terms to a combined Barabaig–Turu influence. They are basically Barabaig, but with Turu modifications, thus suggesting the Turu as intermediaries. The movement of livestock from Turu to Sandawe has been attested to over and over again by informants from both tribes. Nevertheless, the Barabaig have obviously played more than a passive role. Though the two peoples are pretty thoroughly separated at present by a belt of tsetse-infested country stretching across northern Usandawe and southern Mbulu district, this is a fairly recent occurrence, dating only from the 1920's. Previously, the Barabaig were fairly frequent visitors to Usandawe, searching for pasturage and also raiding, while the Sandawe traveled into Barabaig country to collect salt at Lake Balangida Lelu. Sandawe oral literature is replete with references to the Barabaig, particularly as regards warfare. One such story from the Barabaig view has already been cited. Futhermore, some Barabaig took up residence in Usandawe, forming what is now the *Alagwa* clan of the Sandawe. This is confirmed by their own clan history and by histories of other clans as well.[12] Eventually they came to occupy a dominant position among the Sandawe, mainly because of their ownership of livestock and consequent capability for purchasing wives, and for their reputed powers as rainmakers.[13] The whole process evidently took the form of a gradual infusion rather than a violent invasion. As the Alagwa spread livestock to the Sandawe, they themselves became assimilated into the larger society, and the women they took passed on the Sandawe language and customs to their children. When the German administration arrived, the Alagwa considered themselves Sandawe, and due to their high status the Germans formalized them as leaders of the tribe.

Once more, the Gogo have to be considered as a possible source. The tradition of the Gogo passing on livestock to the Sandawe is encountered in both groups, being particularly strong, as one would expect, along the contact zone in the southeast. Also, in this area, some Gogo animal terminology creeps into the Sandawe vocabulary. However, few Sandawe in the southeast are stockowners (probably no more than one household in five), and Gogo influence has undoubtedly been minor at best.

The Masai seem to have been even less important. Sandawe oral literature is full of references to the Masai, but all of these are in connection with their raids to steal livestock. Apparently, by the time of the arrival of the Masai, the Sandawe already possessed stock in great enough numbers to be tempting. Furthermore, everywhere they are found in Usandawe, the Masai form their own tight-knit communities. They refrain from any substantial intermixture. If a Sandawe wife is taken, she must become Masai and cut her former ties completely. It is said by the Sandawe that it is impossible for a Sandawe man to marry a Masai woman, a statement readily borne out by the marriage statistics in Table 4.

Up to now, nothing has been said about contacts to the east, with the Rangi and Burunge. All indications are that until quite recent times, these have been very limited. Substantial Rangi movement west from Kondoa town dates from the late 1940's with tsetse clearance in the Bubu Valley, while the Sandawe seem never to have been strongly represented in the Songa Hills or along the eastern bank of the Bubu. The Burunge are an extremely small group in numbers, with the majority of the population settled to the east of the Sandawe border in the vicinity of the present-day Goima. It is doubtful that with such few numbers and with the comparative scarcity of Sandawe in the Farkwa-Tumbakose-Maxorongo area, where contact would have been most likely, that the Burunge could have played much of a role, if any, in the transformation from hunting and gathering to settled husbandry by the Sandawe.

The recent nature of contact with the Rangi and Burunge is revealed by how little the Sandawe know of their customs and, in turn, by how ignorant these two peoples are of the Sandawe. About all that respective informants

FIGURE 19 The administrative center of Kwa Mtoro is the only real village in Usandawe. This overview toward the Songa Hills shows the government facilities in the foreground and the cluster of small shops, mostly Arab-owned, in the center.

could say was that "those people are different from us." In contrast, ask the Turu, Gogo, and Sandawe to talk about one another and they can go on for hours concerning marriage, initiation, eating habits, and so forth. Similarly, there are many Sandawe fluent in the Turu and/or Gogo language, while a considerable number of Turu and Gogo can speak Sandawe, despite its uniqueness and difficulty. I encountered very few Rangi and Burunge who were able to converse in Sandawe, and vice versa; an observation confirmed by many informants.

As Map 6 shows, few Rangi and Burunge have penetrated Usandawe. The sharpness of the ethnographic boundary between the Sandawe and Burunge is particularly striking. There is virtually no intergradation. Rangi are somewhat more common, but most of them are government employees and are therefore concentrated in and around the administrative center of Kwa Mtoro (Figure 19). A few are also found at Farkwa. The Rangi with their progressive and aggressive nature have come to dominate government posts in Kondoa district. Lastly, the frequency of intermarriage between Rangi and Burunge with Sandawe is negligible. The reluctance of Rangi men to marry Sandawe women has already been noted, while there is almost no opportunity for Sandawe men to take Rangi women. The Sandawe say it is impossible for them to get along with the Burunge and that this is the reason why so few intergroup marriages take place. No specific reason for the animosity is given.

In sum, it seems most likely that the spread of crop and animal husbandry practices to the Sandawe has emanated from a dominantly Turu source, with probable subsidiary contributions from the Gogo and Barabaig. After perhaps an initial stimulus from severe famine, it was accomplished primarily through a gradual process of infiltration and assimilation, not sudden invasion, and consequently the Sandawe were able to maintain their separate identity. An essentially similar state of affairs can be witnessed today among the Hadza. According to Woodburn, they are being changed and absorbed by Isanzu, Iraqw, and Sukuma migrants who are encroaching upon traditional Hadza territory.[14] The Isanzu, in particular, are taking Hadza wives, but the Hadza tradition still survives in the children. These "hybrid" groups, followed by a few "pure" Hadza, generally settle close to Isanzu communities and have begun to cultivate regularly.[15]

The conversion away from hunting and gathering was undoubtedly reinforced by a consequent growth in population due to in-migration and the greater numbers able to be supported by agriculture. Under continued growth, the availability of game animals and bush products would necessarily decline both relatively and absolutely, further strengthening the hand of a more domesticated existence. How many people a strictly hunting and gathering economy could support cannot be exactly measured, but it is unlikely that the total exceeds 1,000 within the confines of present-day Usandawe. Woodburn has recorded an average land requirement figure of 2 sq mi per person among the Hadza,[16] where the natural potential for "wild" food products, especially game, is quite similar to that for Usandawe. This is actually the most favorable ratio yet cited for hunters and gatherers, others running even as high as 100 sq mi per person.[17]

Without archeological investigation, another indeterminate dimension is time. Exactly when the transition of the Sandawe began remains pretty much a mystery. If Wilson's reckoning on the time of expansion for the Barabaig is correct, then the adoption of livestock would fall sometime within the last 250 years.[18] Just prior to this, he claims that the parent Tatog were living around Mount Elgon in Kenya and had not yet begun their southward expansion. Hunter has reckoned the Turu to have reached the Singida area about 200 years ago,[19] but the more thorough works already cited by von Sick and Schneider offer no estimate of arrival. We can probably assume that the initial adoption of crop cultivation took place about 200 years ago, if it is remembered that this is at best a rough approximation and likely will be altered by future investigations.

So far we have been attempting to trace those probable changes that took place before documentation began in the 1890's. Now we will proceed with what has transpired in Sandawe subsistence practices over the past 60 to 70 years.

CHAPTER 5

RECENT CHANGE

More recent modifications in Sandawe subsistence have led to an increasing predominance of crop cultivation and livestock herding. Several new additions have been made to the agricultural system, while at the same time hunting and gathering continue to fade farther and farther into the background. Indeed, one can see the final death blows being dealt to the old tradition. An increasing proportion of the younger generation knows little, and in some cases virtually nothing, about hunting skills and expresses little desire to learn them. They frequently confuse animal names, and many have never seen the likes of koodoo, bushbuck, eland, wildebeest, and hartebeest. Several old informants sincerely expressed to me their concern over the lack of bush knowledge by the young, but admitted that they themselves now hunt on relatively infrequent occasions. The basic reasons behind the demise are not too hard to seek out.

Without question, increasing population pressure has played a continuing role in the decline of hunting. A fairly obvious generalization, proven over and over again in archeological records and in recent documented history, is that the greater number of human beings occupying a given area the fewer will be the number of wild animals. The animals simply cannot compete as their orbit of movement becomes circumscribed, as the available browse and forage lessens, and as the hunting becomes more intense.

The first population figure we have for Usandawe, other than pure guesswork, comes from Dempwolff, who quotes 16,270, based on an assessment for taxation in 1907–08.[1] He did not stipulate whether this referred to just Sandawe or the total for the chiefdom, but presumably it was the latter since the next census in 1921 showed 13,852 Sandawe for the whole of Kondoa district.[2] Unfortunately, it was not possible to locate any further areal refinement, so the exact number of Sandawe and other residents in Usandawe remains indeterminate. In 1928 another count was taken, this time allowing us to speak with somewhat greater precision. The total number of Sandawe in Kondoa district jumped to 20,092 with 19,424 being located within the chiefdom boundaries.[3] Together with 3,885 Turu and 2,231 others, Usandawe carried a population of 25,540. A complete recount of the district was performed in 1931 because of inaccuracies discovered in enumerating the Rangi and Burunge. By 1931, the recorded figure for Usandawe reached 28,197, of which 21,588 were listed as Sandawe.[4] Undoubtedly, a good share of this rather substantial increase in the space of one decade has to be attributed to more accurate counting, but still there must have been considerable natural increase as well.

With the onset of the Great Depression, which was followed closely by World War II, the next population census was delayed until 1948. Interestingly enough, the statistics for Usandawe were virtually the same as for 1931, with 21,202 Sandawe and 6,691 others, of whom 3,553 were Turu, making a grand total of 28,163.[5] Throughout Kondoa district there were 22,286 Sandawe and 27,699 in the country as a whole. Some young men were now beginning to trek to the plantations and commercial estates around Arusha and Tanga for wage employment. In the last official census of 1957, the trend toward out-migration and actual depopulation hinted at in the 1948 figures were intensified. A total of 28,309 Sandawe were recorded in Tanzania, but only 20,607 were found in Kondoa district and 19,607 in Usandawe.[6] The total population for the chiefdom dropped to 26,868.

Despite a recent overall population decline, the area in and immediately adjacent to the Sandawe Hills has at least held its own and probably has even witnessed a slight increase. Present densities around Ovada and Kurio-Kwa Mtoro reach in excess of 100 persons per sq mi, while much of the remainder of the hills has general densities of over 20 persons per sq mi. The reasons for such concentration will be analyzed somewhat later. For the discussion at hand, it means that the only wild animals found in any number are dik-dik, duiker, bush pig, porcupine, hyrax, and hares. Large game, with the exception of migrating elephants, is almost entirely confined to the margins of Sandawe settlements: in the Songa Hills, on the Kioli *mbuga,* in the eastern Bubu and Mponde valleys, and on the high plateau in the far north. Children are no longer raised in intimate contact with game animals, and at least several days are required to make a successful expedition to the major hunting grounds. More and more men are reaching the conclusion that it is just not worth the time, effort, and danger involved.

The danger does not include only immediate physical harm, but also a fear of being caught breaking one of the game laws. The Germans began to enforce hunting regulations to help conserve the fauna and as a means of forcing the Sandawe to settle down, and the pressures have continued and increased up to the present time. For example, licenses with quotas attached must be purchased; certain animals like giraffe are entirely excluded; nets are forbidden except for bush pigs, baboons, and other pests; poisoned arrows and pits are illegal; no hunting can be done from blinds; and a 500-yd restriction is placed around watering points.[7] It is openly admitted by Sandawe hunters that they fear the imposition of fines and possible imprisonment.

Two closely interrelated factors that have greatly affected the status of hunting are the spread of Christianity and education. There are three Catholic mission stations in Usandawe—at Kurio, Farkwa, and Ovada—staffed by the Italy-based Passionist Fathers. Kurio was first opened in 1909, followed by Farkwa in 1929 and Ovada in 1939, and all together between one-half and two-thirds of the Sandawe presently would at least claim to be Christians. A few Muslims also live around the main trading centers. Though the missions have not preached against hunting, they have discouraged continuance of the circumcision camps and consequently some young boys from the most ardent Christian families no longer participate in that important source of training. At present, the missions sponsor seven lower primary schools, grades 1 through 4; two extended primary schools, grades 1 through 6; and a middle school in Usandawe, while the government runs three lower primary schools. Total enrollment is roughly 1,200 students. As a result of increasing education both parochial and public, hunting has become regarded as an expression of the primitive past and agriculture as representing the progressive future. To be civilized, one must till the soil. A feeling of shame, and, by a few, even disgust, is indicated at the eating of such foods as baboons, monkeys, jackals, cats, snakes, and rodents, and only with great difficulty will many of the "moderns" admit that their "uncivilized" or "pagan" cohorts still pursue these animals. Less delectable items, like the ground hornbill (*Bucorvus leadbeateri*) and civet, both of which have rather strong-smelling flesh, are being rejected. The Sandawe are beginning to be choosy about what they eat.

The products of gathering have not followed quite so precipitous a decline. Some of the larvae, caterpillars, and locusts have an "unfit" stigma, but honey is still avidly sought after, and most of the fruits, berries, and nuts are consumed when possible. However, the more the bush is cut down and cultivated fields appear, the less will be the availability of the majority of wild foods. Again, it is strictly a matter of competition.

In agriculture, the most significant changes are related to the coming of a cash economy and the provision of famine relief. A secondary contribution has been made by agricultural extension works.

The need to pay taxes and the desire to purchase such accoutrements of civilization as cotton cloth (which completely replaced animal skins for clothing), sugar, bicycles, transistor radios, etc., have led to the growing of two cash crops, groundnuts and castor. Groundnuts were introduced by the British administration shortly after World War I in the hopes of finding a viable money-producing crop for the semiarid regions of the country. Only the Iramba, it seems, grew them on any scale in central Tanzania before this time. Today, for the Sandawe, groundnuts are the principal cash crop and are also used as a relish with *ugali,* supplementing and in some instances replacing Bambarra nuts. All Sandawe who were interviewed claimed to grow them, and, as already indicated, they are planted in separate fields, on the average about one-fourth to one-third of an acre in size per household. The common East African practice of interplanting groundnuts with the grains is not followed by the Sandawe (Figure 20). Castor was first stressed as a cash crop in Usandawe during the 1950's. It is grown in the main grain field, usually mixed up with the other crops, but occasionally separated. Two varieties are found, spined and nonspined; the former is overwhelmingly dominant. On the whole, the Sandawe are not very enthusiastic about castor because of the extra work involved and the low current price. Should it not be sold, some complain, "What can we do with it? It cannot be eaten."

Unfortunately or fortunately, depending on one's point of view, Usandawe has never come under the direct influence of European settlement or large-scale commercial estates. We have seen that some Sandawe have gone off to work where these are found, but an enduring stimulus to a cash economy has not been present. The ideas of money, of buying and selling, have only gradually penetrated the countryside, and concern with

FIGURE 20 Groundnuts and sweet potatoes begin to sprout sometime in March and are harvested normally in June.

these facets of life are not particularly central issues in everyday conversation. As long as taxes can be paid and a few luxuries purchased, there is considered to be enough money.

It is clear, from the writings of the early observers and the communications of later resident agricultural officers, that up until quite recent times maize was a secondary grain, ranking well behind bulrush millet and sorghum. Dempwolff emphasized the dominance of the latter two foods,[8] while a report summarizing agricultural conditions in Tanganyika for 1945 gave the following estimated acreages for the old Central Province of which Usandawe was a part: bulrush millet–383,760, sorghum–118,000, and maize–18,710.[9] In 1950, a sample census of agriculture for the World Agricultural Census revealed essentially the same pattern. For Kondoa district, 3,197 pounds of bulrush millet, 2,661 pounds of sorghum, and 923 pounds of maize were recorded per taxpayer.[10] However, by 1965 maize had assumed a considerably more significant position. In fact, sampling estimates by the local field officer of the Ministry of Agriculture placed it as the number one crop in Usandawe from both the standpoints of acreage and tonnage.[11] This observation was borne out in my dietary statistics, which revealed that maize was eaten as the main food at 56 percent of the meals, compared with 19 percent for sorghum, 16 percent for bulrush millet, 5 percent for sweet potatoes, and 4 percent for groundnuts, Bambarra nuts, and the two legumes.

The beginnings of the transition (perhaps revolution is more appropriate) seem to date from the late 1950's, but the real stimulus apparently came with the severe famine of 1961–62. During this time, surplus maize from the United States was distributed for famine relief, thus averting a major catastrophe, since the virtual failure of the rains throughout central Tanzania caused almost universal crop failure. The Sandawe refer to it as the *Amerikani naragu,* or American famine, because of the imported maize. With the coming of the next planting season, maize seeds were the only ones available in any number, and thus they got planted. Most Sandawe informants agreed that this was when and why they started planting maize as a main grain crop and that they have been increasing its proportion ever since.

There are several other reasons behind the rise in popularity of maize. The Sandawe will invariably say that they prefer it because it is not susceptible to bird damage when mature. Both bulrush millet and sorghum require a considerable amount of effort, usually in the form of children with slingshots accompanied by shouting and other noisemaking, to drive off the birds during April and May, and it is not unknown for local food shortages to occur because of bird depredations. The main pest species are listed in Table 5. It will be noted that the infamous Sudan dioch (*Quelea quelea*) does not appear, even though Usandawe is located near the main breeding region for these birds in Tanzania.[12] I would venture to suggest that out of ignorance of other species, people have been all too hasty in blaming the Sudan dioch whenever bird damage is reported. Another important reason for the success of maize has been the recent establisment of milling machines at each of the mission stations. Maize is much more difficult to grind than either bulrush millet or sorghum using the traditional mortar and pestle method, but the machines compensate for this problem. Formerly, maize was eaten mainly green, as a buffer food before the other crops were ripe, but now it is employed primarily in flour form for making *ugali.* Maize can also enter the economy as a cash crop, and actually has attained the number two position behind groundnuts in this respect. It brings about the same price as castor, with the advantage of being edible if unsold. Furthermore, it can be traded without the need of middlemen. Finally, there seems to be a certain mystique which has attached itself to maize growing, one associated with being a progressive individual, in contrast to bulrush millet and sorghum which are things of the past and therefore belong to the primitives. This is an extremely difficult proposition to document, but the impression is given by the proud manner many informants displayed when asked about how they grew their maize.

Agricultural extension work has never been systematically pursued in Usandawe, primarily, it seems, because of the traditional marginal productive nature of the area and the consistent shortage of funds and qualified staff. Its most notable achievement has been the introduction and consequent spread of groundnuts. Tie-ridging, drainage-ditching, and more effective methods of manuring have been introduced in the vicinities of Kwa Mtoro-Kurio and Ovada, but these techniques have not been part of a long-term program to demonstrate their respective effectiveness. Therefore, they have not been generally adopted by the Sandawe. Sporadic attempts have been made for years to stimulate the growing of cassava as a reserve

TABLE 5 Principal Bird Pests of Usandawe

Common Name	Scientific Name	Damage
Red-billed hornbill	*Tockus erythrorhynchos*	All of these eat the seeds of maize, bulrush millet, sorghum, and eleusine when they are planted, plus the mature heads of bulrush millet, sorghum, and eleusine.
Yellow-billed hornbill	*Tockus flavirostris*	
Speckled pigeon	*Columba guinea*	
Red-eyed dove	*Streptopelia semitorquata*	
Ring-necked dove	*Streptopelia capicola*	
Pink-breasted dove	*Streptopelia lugens*	
Green pigeon	*Treron australis*	
Emerald-spotted wood dove	*Turtur chalcospilos*	
Namaqua dove	*Oena capensis*	
Green-winged pytilia	*Pytilia melba*	
Jameson's firefinch	*Lagonosticta jamesoni*	
Red-billed firefinch	*Lagonosticta senegala*	
Cutthroat	*Amadina fasciata*	
Red-cheeked cordon-bleu	*Uraeqinthus bengalus*	
Streaky seed eater	*Serinus striolatus*	
Helmeted guinea fowl	*Nimida mitrata*	These dominately eat the seeds of newly planted maize, bulrush millet, and sorghum and also occasionally ripened bulrush millet and sorghum.
Kenya crested guinea fowl	*Guttera pucherani*	
Scaly francolin	*Francolinus squamatus*	
Grey-wing francolin	*Francolinus afer*	
Coqui francolin	*Francolinus coqui*	
Crested francolin	*Francolinus sephaena*	
Fischer's straw-tailed whydah	*Vidua fischeri*	These two eat only bulrush millet and sorghum when they are ripe.
Paradise whydah	*Stenganura paradisaea*	

All terminology for birds used in this report comes from John G. Williams, *A Field Guide to the Birds of East and Central Africa* (London: Collins, 1963).

famine food, all of which have failed. Most Sandawe households have grown cassava at one time or another, though it is infrequently seen today. The almost universal complaint is that they have not been shown any effective measure to control the rooting-out done by bush pigs and porcupines. By and large, one is forced to conclude that Usandawe has been rather neglected by the Ministry of Agriculture. This is even more strikingly true when one moves away from the population centers. As of 1966 some of the people in the backlands have never seen even the lowest rank of agricultural field officer.

The livestock sector of the economy has undergone two basic changes since the beginning of the present century. First, there has been a general rise in animal numbers, due primarily to the increasing availability of preventive treatments and cures for such formerly decimating diseases as trypanosomiasis, blackleg, and anthrax, plus the provision of several bore holes to provide year-round adequate water supplies. Table 6 gives the livestock counts for years when comparatively reliable data are available. It can be seen that before the present, data are scarce. Second, monthly livestock markets have recently come into existence at Kwa Mtoro and Farkwa. They provide a regular opportunity to buy and sell animals and are accompanied by the usual coterie of merchants selling pots and pans, sandals, and what have you. Both markets are still fairly small when compared to those held among the Gogo, Turu, and Rangi, and, in fact, a good portion of the active participants are Masai.

TABLE 6 Livestock Numbers in Usandawe for Selected Years

Year	Cattle	Goats	Sheep	Donkeys
1921[a]	31,501	15,055		
1927[a]	58,254	40,134	8,421	652
ca. 1954[b]		108,371 for all animals		
1961[a]	66,200	46,402	12,134	1,403
1962[a]	69,272	53,586	14,010	1,475
1963[a]	73,469	54,536	15,150	1,855
1964[a]	75,202	45,179	12,007	1,052

[a]From Local District Council tax rolls as given in the District Administrative Book.
[b]From the "Central Province-Agricultural Policy," Ministry of Agriculture, Forests and Wildlife, Tanganyika, 1954–55. (Mimeographed.)

In a broad sense, then, the forces bringing change to Sandawe subsistence for the past three-quarters of a century have followed the same pattern as previously. Population growth has reinforced the reliance on crop and animal husbandry as land became more and more scarce. Lately, however, increments to the population have come about mainly in response to internal expansion, with little in the way of in-migration, a factor undoubtedly responsible for much past growth, and actually, as we have seen, recent decades have even witnessed a slight out-migration. The last movements in any appreciable number of outsiders into Usandawe were those of the Nyamwezi and Kimbu who came sometime shortly before the arrival of the Germans in the 1890's. Usandawe had ceased to be an obvious population vacuum, and, of course, it was the policy of both the German and British administrations to crystallize tribal boundaries and prevent large-scale unplanned migrations of people. Culture contact has again been all-important, but the actual contact situation itself has undergone transition. Change is now more purposeful and formal, its primary agents being the government and the money economy. It is the overt purpose of both to create wants and needs and then to channel these toward desired objectives. It is rather doubtful that the Turu, Barabaig, Gogo, and Nyamwezi operated in this fashion. Indeed, there is no evidence to suggest that any of the Sandawe's neighbors have played a direct role in recent subsistence changes, though government officials and others have used them to a limited extent as examples. A decline in influence from these directions is understandable, given the Sandawe's previous adoption of many of their neighbor's practices. The gap is no longer so great.

With the background of past changes in hand, the time has come to turn our attention to the task of assessing the potentialities for future modifications in Sandawe subsistence. Given what has gone before, what are the most probable courses for change to take? In order to answer this question, it will be necessary to measure more precisely the physical and cultural resources available to support various alternative strategies.

III

SUBSISTENCE POTENTIAL

CHAPTER 6

CROP PRODUCTION

For several reasons, it would seem most fitting to begin the discussion of subsistence potential in Usandawe with a look at the problems and prospects of crop production. First, we have seen that the major direction of past change has been toward a greater reliance on the products of crop cultivation; consequently, it behooves us to try to determine what further developments can realistically be expected. Second, it was pointed out in Chapter 1 how important agriculture is to the economics of most non-Western countries, particularly to those that, like Tanzania, lack extensive mineral and timber resources and prospects for large-scale industrialization. This importance has long been recognized, if not systematically explored, for, according to one person acquainted with British policy, "The development of Tanganyika depended on the development of its agricultural potential, and every effort had to be made to encourage the production of crops for export in addition to the production of sufficient food crops to feed the people."[1] In the recent Five-Year Plan for the country, agriculture ranks second to infrastructure in expenditure, with 27.1 percent of the development funds to be allocated for its improvement.[2] The Sandawe and the administration appear to be thinking along the same general lines, and we should, therefore, attempt to follow their priorities.

"The ordinary flowering plant lives its whole independent life with its roots in the soil and its remaining parts exposed to the atmosphere. . . ."[3] Simple as the statement may be, it eloquently emphasizes the preponderant influences exerted by climatic and edaphic conditions on the life cycles of angiosperms. They are the two environmental factors that, of necessity, must be taken into detailed consideration whenever the discussion turns to crop production and development. Without an adequate knowledge of each, there is very little that one can meaningfully say about the subject.

WATER BALANCE

Within the broad realm of climate, the most significant element for crop growth is moisture availability. This is particularly true in the tropics where, except under unusual circumstances, light, temperature, and wind cannot be considered as limiting factors. The broad outlines of the rainfall regime in Usandawe have already been discussed, but only in very general form; moreover, income is only one-half the picture. In order to complete it, we will have to bring in an expression of moisture outgo, or more precisely evapotranspiration. Two calculations are possible, actual evapotranspiration and potential evapotranspiration. The latter is considered preferable because it can be determined without expensive and time-consuming instrumental recordings and because it gives an estimation of optimum plant growth under conditions where the moisture supply is limited. How much moisture is needed for the theoretically maximum crop production? The derived figure can then be compared with moisture availability to see how closely conditions fit the ideal.

A voluminous recent literature exists on the measurement of potential evaporation and/or evapotranspiration.[4] The two basic methods are the use of water-filled pans for recording rates directly and the use of empirical formulae that attempt to estimate rates in terms of other climatic variables. On appearance, the former would seem to be the more logical and accurate approach, though in practice a number of difficulties have arisen that cast doubt on its usefulness. These include variations between pans in thermal capacity, exposure, and color, plus the clarity of water, and such factors as leaks and splash.[5] Of the various empirical formulae, the one devised by Penman has the most support from agronomists in the tropics, reportedly because it makes fewer assumptions than the others.[6] It is employed by the East African Agricultural and Forestry Research Organization and therefore is used in the forthcoming analysis.

The meteorological data required for the computation of the Penman estimate include mean run of the wind in miles per day, mean daily temperatures of the air and of the dew point, and radiation in calories per square centimeter per day or mean hours of bright sunshine. There is no station in Usandawe recording the above information. However, one is close by, in Burunge country, where the Bubu River swings westward around the southern end of the Songa Hills. Called Gwandi, the station has been operated under the auspices of the Water Development and Irrigation Department since 1960. Its elevation is about 4,000 ft, somewhat lower than the mean for Usandawe, but not appreciably so, and consequently it should serve as a fairly representative approximation.

The precise theory and exact form of the equation will not be set forth here. Those readers who are interested in the theory can consult the 1948 and 1963 publications by Penman,[7] while the calculation procedures are derived from an article by McCulloch that conveniently reduces them to tabular form once the meteorological data are known.[8] When these calculations are carried out, we have an estimate of E_o, or open water evaporation. To compute to E_t, evapotranspiration, a factor E_t/E_o is required. The factor apparently varies to some extent with the type of vegetative surface and its stage of growth, but 0.8 has been generally accepted as the standard yearly coefficient for tropical areas.[9] We now have all we need to make some general statements about the water balance in Usandawe.

Given the long dry season from roughly May through November in Usandawe, one would expect to find a yearly moisture deficit, a situation realized in the statistics. The total potential evapotranspiration demand calculates out to 58.4 in., leaving a balance of –34.6 in. at Farkwa and –33.2 in. at Kurio. These figures fall within Thornthwaite's limits for semiarid climates.[10] However, such a statement is useful only at a very broad level of analysis. It has little direct agronomic significance. It is much more pertinent to restrict attention to that portion of the year, namely the growing season, when plants have the possibility for actively transpiring. For Usandawe, the relevant time span can be taken as occurring from the end of November through April.

For the growing season as a whole, there is still an average moisture deficiency. Evapotranspiration demand is 25.7 in., while precipitation at Farkwa is 22.1 in. and at Kurio, 23.7 in. However, since the growing season itself can also hide important internal variations, even a further refinement is desirable. The month is probably the most widely used unit, though the 10-day period is becoming increasingly popular because of its ability to detect short-term fluctuations that often have considerable significance for plant growth. Table 7 presents a breakdown of the growing season in Usandawe employing two 10-day periods and a residual for each month. The figures demonstrate the precarious moisture balance situation prevailing at every stage of the season. Farkwa has only 3 out of 16 periods when there is an average income exceeding outgo. Kurio looks somewhat better, with 9 surplus periods, but in each case the balance is slight, and, given the runoff that will naturally occur in hill country under conditions of tropical downpour, it is unlikely that one could reasonably expect a positive balance of moisture availability over requirements at any time. There is no reliable "wet" period.

From the point of view of agricultural risk, a much more meaningful analysis can be made using statements of probability. According to Curry's apt phrasing, "policy must be based on an economic assessment of the probabilities of risks and gains: a strategy, as in a game of chance with nature as the opponent, is implicit in a farm program."[11] In line with this notion, the yearly variations of precipitation for each trimonthly period during the growing season at Farkwa and Kurio have been plotted against expected potential evapotranspiration (see Graphs 1–18, pp. 108–125). These reveal that the chances of meeting total moisture requirements during the growing season are only 20 percent at Farkwa and 35 percent at Kurio and that there is no single trimonthly period at either station showing a probability of greater than 50 percent. The onset of the rains is especially unreliable, with probabilities of 15–17 percent for the last 10 days of November. In fact, there is only a 50–50 chance of getting just a light shower at this time. The figures increase rapidly during December, reaching 44 percent at Farkwa and 49 percent at Kurio during the final period of the month. Probability remains above 40 percent in the first two-thirds of January, but then declines through March I, when Farkwa is at 21–22 percent and Kurio 37–38 percent. A secondary rise is recorded for the remainder of March—a 47 percent probability is reached at Kurio for March II and a 40 percent probability at Farkwa for March III—followed by a steady tapering off in April to about 24 percent and 16 percent respectively by the end of the month.

Thus, in the majority of years one can expect for any given time throughout the growing season a moisture income below that required for optimum plant growth conditions. The probability figures amply testify to the precarious nature and great risk of crop production in Usandawe without irrigation, but they should not be taken to imply a regularity of crop failure. We know this is not the case, and it is extremely doubtful whether any

TABLE 7 Average Potential Evapotranspiration (PET) and Precipitation for Usandawe during the Growing Season

Period	PET	Farkwa			Kurio		
		Precipitation	Residual	Cumulative Total	Precipitation	Residual	Cumulative Total
Nov. III	1.81	1.15	–0.66	–0.66	0.84	–0.97	–0.97
Dec. I	1.73	1.24	–0.49	–1.15	0.90	–0.83	–1.80
Dec. II	1.51	1.40	–0.11	–1.26	1.61	+0.10	–1.70
Dec. III	1.59	1.59	0.00	–1.26	1.86	+0.27	–1.43
Jan. I	1.51	1.66	+0.15	–1.11	1.75	+0.24	–1.19
Jan. II	1.61	1.90	+0.29	–0.82	1.63	+0.02	–1.17
Jan. III	1.89	1.48	–0.41	–1.23	1.92	+0.03	–1.14
Feb. I	1.60	1.62	+0.02	–1.21	1.84	+0.24	–0.90
Feb. II	1.72	1.53	–0.19	–1.40	1.49	–0.23	–1.13
Feb. III	1.29	1.13	–0.16	–1.56	1.53	+0.24	–0.89
Mar. I	1.63	1.11	–0.52	–2.08	1.75	+0.12	–0.77
Mar. II	1.69	1.42	–0.27	–2.35	2.02	+0.33	–0.44
Mar. III	1.77	1.39	–0.38	–2.73	1.60	–0.17	–0.61
Apr. I	1.43	1.43	0.00	–2.73	1.09	–0.34	–0.95
Apr. II	1.43	1.18	–0.25	–2.98	1.08	–0.35	–1.30
Apr. III	1.45	0.84	–0.61	–3.51	0.79	–0.66	–1.96

group of people would depend on crops for their livelihood in an area where failure is more frequent than success. It must be remembered that potential evapotranspiration is a measure of optimum need and there is apparently some range of variation between specific crops in how closely they must approach this figure during each stage of their growth cycles in order to produce at least up to a basic subsistence level.[12] Unfortunately, detailed specifics on the requirements in East Africa for various crops are not available. Nevertheless, it is still possible to make some general observations on future suitability.

Of the staple small grains, there seems little question that sorghum and bulrush millet are more adaptable to areas of low and erratic rainfall than is maize. In trials carried out by EAAFRO, sorghum consistently yielded higher than maize under semiarid conditions.[13] Experience in Usandawe suggests that sorghum will produce sufficiently for local needs on as little as 16 to 18 in. of precipitation during the growing season. But its greatest attribute is the ability to weather extended periods of little or no rainfall. Should the rains break late, the seed can remain dormant in the soil until there is sufficient moisture to germinate, while it has the added capacity of being able to recover from wilting once the growth process is under way. Bulrush millet is able to get by on even a little less total moisture than sorghum and also can remain in dry soil for a considerable length of time, but it does not have the capacity of reversing the wilting process.[14] Maize has neither the ability for tolerating such low total moisture income nor of surviving extended periods of drought once it is in the ground. At the very minimum, 20 in. of rainfall are required, while more than 20 successive days with little or no moisture will in most cases administer the *coup de grâce.*[15] These various requirements were strikingly illustrated in the Farkwa area during 1964–65 and 1965–66. In the former year, the grains were planted at the end of November as usual, but no rain was recorded until December 17. The maize did not germinate, and there was a scurry to replant when the rains came in earnest in the first part of January. Not everyone made it on time, and localized food shortages resulted during the following dry season. Both sorghum and bulrush millet reportedly yielded normally. In 1965–66, there were fairly good rains in December, enough at least to start crop growth, but then very little fell in all of January. The maize wilted and died, in contrast to the sorghum and some millet that endured until the advent of heavy rains in late February through March (Figure 21). With their newfound enthusiasm for maize, the Sandawe seem to be increasing rather than decreasing agricultural risk.

For the comparative performance of other food crops we have to rely

FIGURE 21 Due to erratic early rains, this maize field at Farkwa is in an advanced state of wilt.

almost entirely on local accumulated experience, since relevant research findings are virtually nonexistent. On the evidence of oral testimony, Bambarra nuts, cowpeas, and the various cucurbits grow all right in most years, though cowpeas are notoriously low yielders. As has already been mentioned, sweet potatoes and haricot beans fail quite frequently due to poor timing in planting. It would seem that unless the fields are prepared and planted by the end of February so as to take advantage of the March upswing in precipitation, there is little hope of consistent success. Without the pest problem, cassava could be grown more widely. Though rainfall is far from ideal, the crop will yield satisfactorily in most years. Eleusine does not do well both because of the low and erratic rainfall and because it requires somewhat cooler temperatures during its growth than are achieved in Usandawe, where daily maxima normally run into the high 80's and low 90's. In any event, eleusine is used almost exclusively by the Sandawe for making *pombe.* It is rarely eaten in porridge form.

Ruthenberg has compiled a list of cash crops that he considers could make significant contributions to the economic development of Tanzania.[16] He includes sisal, tobacco, cotton, coffee, tea, pyrethrum, cashew nuts, sugar cane, wheat, rice, cassava, cocoa, kenaf, castor, coconuts, groundnuts, onions, rubber, oil palm, sesame, soya, seed pulses, and maize. Of these, all but tobacco, wheat, cassava, castor, groundnuts, maize, and cotton can be immediately eliminated because they have much greater precipitation demands than can usually be met in Usandawe, while maize and probably also cotton are distinctly marginal. As other considerations are brought to bear, we shall see the list of potential crops dwindle.

OCCURRENCES OF FAMINE

The focus so far has been on average or expected moisture conditions, but Usandawe is an area where the unusual and unexpected can strike with great severity. These are times when famine sweeps the countryside, when food shortages are general rather than local. We have witnessed how famine has apparently had a profound effect on Sandawe history, and its incidence will undoubtedly play a role in shaping future events. In recorded times, that is, since the 1890's, five major famines have visited the countryside: 1917–19, 1943–45, 1948–50, 1953–55, and 1961–63. The last four are documented in the rainfall records of Farkwa and Kurio, and it will be instructive to look at these briefly to see what great risks crop production runs in Usandawe.

1943–45 The 2-year famine of 1943–45 had its origins in the rainy season of 1942–43. During December 1942, Kurio reported 4.2 in. of precipitation for the first 20 days of the month, but then 32 completely rainless days followed. Violent storms broke at the end of January, 7.0 in. of rain falling in 6 days. Unfortunately, only light and intermittent showers were registered for the remainder of the growing season. Just two of the remaining trimonthly periods had a total moisture income greater than an inch, while another long rainless stretch of 21 days occurred in the middle of March. Farkwa did not come into being as a recording station until January. Its records reveal a similar pattern to that of Kurio, with virtually no rain falling in March and April.

Bad conditions continued into the next season, especially at Farkwa. The early rains failed completely; both stations recorded only 0.1 in. to the last day of December. On December 31, a heavy storm dropped 2.2 in. Kurio continued to experience fairly regular rainfall from this time onward to the end of April and consequently widespread crop loss was avoided. Farkwa was not so fortunate. After the storm of December 31, rainfall was fairly light up through most of March. Four of eight 10-day periods had less than 1 in. and none reached potential evapotranspiration demand. Nearly 2 in. of rainfall came during the last part of March, followed by 8.4 in. in the first 10 days of April and a total of 13.2 in. for the month, the highest on record for the station. It was a case of "too much too late," and crop failure was virtually complete for the second straight year.

1948–50 The growing season of 1947–48, which started off the severe food shortages experienced in 1948–50, began as if it were going to be a good year. Kurio received 3.9 in. in the first 10 days of December and Farkwa 5.0 in., with two light showers in November. The encouraging start, however, was short-lived. At Kurio, there followed 15 consecutive rainless days during the middle of December, and rainfall from then until the end of February totaled 7.7 in. The situation picked up suddenly and 8.8 in. fell within the first 20 days of March. Then just as suddenly it diminished, and for the remaining 40 days of the growing season the moisture income totaled 1.5 in. Farkwa maintained a better early position, getting 8.1 in. in January, but witnessed the same drastic reduction in March and April. Complete crop failure was not reported, though yields were considerably reduced, and thus food stocks began running low toward the end of 1948. There was certainly no local surplus to meet the extreme conditions that were shortly to develop.

Again in December 1948, as in the previous year, there was cause for optimism in Usandawe. Though no rain whatever fell until the middle of the month, by the first of the year Kurio had a total of 3.2 in. and Farkwa 4.6 in., enough to get the crops well under way. Then disaster struck. Over the remaining 4 months only 7.6 in. of rain fell at Farkwa, making a total of 12.2 in. for the entire growing season. During one stretch there were 23 days in a row in January without any rain at all, and another period from March 7 through April 10 when the total was a meager 1.2 in. As would be expected, crop failure was virtually complete throughout southern and eastern Usandawe. In the west, things were only slightly better. Kurio managed 16.9 in. of precipitation for the growing season, but again there were the two periods of almost no rain in January and March–April. Some sorghum and bulrush millet apparently survived, but enough for two or three month's food supply at best.

1953–55 Three years later, drought and subsequent famine struck once more. No rain came to Kurio for the 1952–53 season until January 1, and after 13.3 in. had accumulated through February 4, a pronounced dry spell set in. From the remainder of February to March 22 only 0.5 in. fell, including a stretch of 34 straight rainless days. The 3.5 in. received from March 22 until the end of April salvaged some sorghum but little else. The same pattern prevailed at Farkwa, with the long dry spell in February and March managing even an extra four days. Next year conditions deteriorated further, as the rains failed for the entire duration of the growing season. The most intense drought again came right in the middle, from the end of January through March. For this period in excess of two months, Kurio had 3.8 in. of precipitation and Farkwa, 2.4 in. Totals over the entire growing season amounted to 15.2 in. and 12.2 in., respectively.

1961–63 The final severe famine of recent times began with the failure of the early rains of 1960–61. At both stations nothing but sporadic showers were recorded up to the very end of January. During one stretch, Kurio experienced 33 days with a total income of 0.3 in., while Farkwa got the same amount in a 29-day stretch. Conditions improved somewhat from the end of January onward, but did not remain consistent enough to salvage the situation. For instance, Kurio received only 1.9 in. from the last 10 days of March to the beginning of May, and Farkwa in 27 days from the middle of February to the middle of March managed a mere 0.9 in. In contrast, the next season the rains broke in wild abandon. Through January of 1962, 24.6 in. were recorded at Farkwa and 23.5 in. at Kurio. Weeds proliferated, and so intense and frequent were the storms that weeding operations were seriously delayed. Furthermore, much of this water went to runoff and was not available to meet the deficits that soon appeared. Kurio, from February through March, saw the rainfall drop off to 1.4 in., with another period during the final three weeks of April managing 0.5 in. Farkwa had only 5.8 in. from the last 10 days of February to the end of the growing season. Once more the food supply would not see the people to the next harvest.

Given reliance solely on atmospheric moisture for meeting requirements, periodic crop failures will remain part of the system in Usandawe. Whether four droughts in less than 20 years is the expected frequency and whether serious food shortages run in 2-year clusters remain to be seen. Not enough is known yet of synoptic conditions in East Africa to predict with any great reliability, though the statistics in Table 7 and the discussions of the above famine periods suggest that a tendency toward a bimodal pattern of rainfall might provide part of the answer. Most years show a tapering off of the rains in February or March before a second upswing begins. On an average basis, Farkwa reaches a new low during March I and Kurio during February II. However, averages conceal the real magnitude of the decline, since there is little consistency in timing from year to year, and it would seem that the duration and intensity of the dry spell go a long way in determining whether or not success is achieved.

In any event, one can hardly question the need to establish some effective buffers to absorb these shocks. As it stands now, the Sandawe have no devices of their own that they can employ with certainty. The availability of game and bush products is too limited to be of any substantial assistance. Besides, when the rains fail, many of the bush fruits do likewise, while the animals will be in poor shape, if they have not already moved outside Usandawe in their search for water and forage. During the famines of 1943–45 and 1948–50, deaths that could be directly attributable to lack of food were not uncommon, particularly among the old and very young. Both occasions are remembered by the Sandawe as times when many of the men went off to northern Irangi, Gorowa country, the Mbulu Highlands, and the Iramba plateau in search of food. Not much success was encountered, since these areas were experiencing similar difficult times. It is also reported that some Nyamwezi traders came through buying children for grain and cash.

Virtually no famine deaths were reported for 1953–55 and 1961–63. A new factor had entered the Sandawe economy, famine relief. In the former period it was obtained from other areas in East Africa: from Sukumaland, from Buganda, from the Kenya Highlands, and from the Southern Highlands of Tanzania. The same sources could not be used in 1961–63, for nearly all of East Africa was in the grips of drought and food shortage. The relief came instead, as was mentioned previously, from American surplus maize. The immediate benefits of such relief are self-evident, but the long-term effects are more difficult to assess. Some Sandawe expressed the view that there was little reason to worry about whether the crops grew or not because now food always came if needed. To them there is no more problem of risk. The same general response has been noted elsewhere,[17] and it is one that is difficult to reconcile with the need for agricultural improvement.

A form of local famine assistance was attempted after the problems of 1943–45, which, in theory at least, holds some promise of success. During the remainder of the decade, concrete grain-storage bins were constructed at population centers all across central Tanzania. In Usandawe they were set up at Kwa Mtoro, Farkwa, Lalta, Sanzawa, and Tumbakose. The idea was that in good years the local people would contribute a portion of their grain harvest and later draw on the surpluses when shortages developed. Two problems arose that resulted in the scheme's failure. First, most people were highly suspicious about putting what they had grown into the bins and not being assured of an equal quantity when the time came for withdrawal. Second, the bins were not large enough to hold sufficient grain to meet large-scale food shortages, as was painfully obvious in 1948–50 and again in 1953–55. Today, derelict bins, empty except when outside famine relief is brought in, are a permanent feature of the landscape and a constant reminder of the problems of erratic rainfall.

SURFACE-WATER AND GROUNDWATER RESOURCES

An extremely important consideration in the assessment of crop potential is the availability of surface-water and groundwater supplies. Can irrigation alleviate the precipitation problem and introduce a new factor into the agricultural system? With just such a question in mind, a survey was begun of the Bubu River during the late 1950's and early 1960's. Three gauging stations were established: at Serya, at Gwandi, and near the junction of the Bubu and the Mkinke rivers. A soil survey was also carried out. The project was never completed and no final report was published. But the raw data are available, and with them we can attempt to draw some tentative conclusions.

Graphs 19, 20, and 21 portray for the Bubu-Mkinke station the depth of water in the channel for the years 1958–59, 1960–61, and 1962–63. What stands out is the extreme fluctuation between any two years and within a single year. As is to be expected, the irregular pattern reflects rainfall in Usandawe and in the Mbulu and Irangi highlands farther north where the river originates. Near-average rainfall occurred in 1958–59 and 1962–63 in the region, while 1960–61 represents a period of serious drought. Obviously, no relief can be expected from the Bubu when drought conditions prevail, the time when it is needed most urgently. Even during comparatively good years, irrigation waters would augment production only to a small degree. Dams can be built to hold the peak flows, but the problem is that these peaks are of such short duration. The channel at this point is only about 15 ft deep and 10 to 12 ft wide and thus, even if it were full, the volume would not be very substantial. At most, a few acres on each side could be regularly irrigated, and even then pumping would be required. Furthermore, there is no hope of extending the growing season with impounded water. Thus, the problem of water supply from the Bubu is twofold: the amount is limited and it is not available when most required (Figure 22).

As if these obstacles were not enough, there are several other factors that also militate against the Bubu's waters making a contribution to crop development. In the area between the Songa and Sandawe hills where, given a water supply, the level topography would be conducive to irrigation, the soils were found to have serious drainage problems due to the previously mentioned high clay content and the widespread presence of a subsurface hardpan layer. Should irrigation water be applied, a substantial investment to improve soil drainage would have to be made.[18] Where it cuts through the Sandawe Hills, the river flows in a deep gorge formed by the Bubu fault, leaving the water generally 100 ft or so below the level of

FIGURE 22 The Bubu River contains only scattered pools of water during the dry season. The sands, however, are moist a few inches below the surface and shallow wells can be dug.

irrigable land. Soils with impeded drainage are once more encountered where the river emerges onto the plains south of Gonga.

The other three rivers of Usandawe—the Zozo, Mkinke, and Mponde—show even less irrigation potential. The first two have very shallow and narrow channels, and for the most part seldom hold enough water to maintain a noticeable flow. The Mponde is larger, but its volume appears to be less than half that of the Bubu, according to the records of a gauging station near where the river enters the Bahi swamp.

Groundwater in Usandawe can be found at four general sites: intrusive dykes such as pegmatite; rock deformations, particularly faults and joints; depressions filled with sediment; and dry stream courses.[19] The latter have been thoroughly explored and utilized by the Sandawe and can be excluded as a potential source for irrigation water, since at best they yield only enough for year-round domestic purposes. Small quantities of water for human and animal consumption are also obtained from shallow wells dug in sediments (Figure 23) and at the break in slope between the steep upper hillsides and the pediment slope where faulting frequently occurs. When there is seepage onto the surface someone will usually plant a little sugar cane and perhaps a banana stool or two. However, seepages are so sporadic and limited in size as to benefit no more than a few dozen homesteads over the entire countryside. The several boreholes that have been dug indicate a potential that does not extend beyond supporting small gardens in their immediate vicinities. Table 8 shows how limited the flows are and also that there is a tendency toward fairly high salt contents.

FIGURE 23 Most of the water for domestic purposes comes from wells dug in the sediment along drainage courses in the Sandawe Hills.

EDAPHIC CONDITIONS

Within the limits imposed by the moisture supply and in light of the fact that there are not markedly perceptible differences in its occurrence throughout Usandawe, one must turn to soils for additional refinement of both present and potential patterns of crop distribution. Soils are the most significant natural resources that can be manipulated for production specialization and consequently, it is hoped, a maximization of returns. Like about every other African people so far studied the Sandawe are cogni-

TABLE 8 Borehole Characteristics in Usandawe

Location	Year Drilled	Depth of Water	pH	Total Solids (parts per million)	Yield (gal/hr)
Poro	1955	136	7.0	1,550	3,000
Farkwa	1956	15	7.5	1,700	2,500
Kwa Mtoro	1957	150	?	1,734	161
Maxorongo	1958	151	8.3	900	100
Kurio	1960	45	7.9	2,214	200

Data obtained from records of the Water Development Division, Ministry of Lands, Settlement and Water Development, Dodoma, Tanzania.

zant of variations in soil fertility and have reacted by constructing a system of land-use preferences. Actually, the system is not based strictly on soil characteristics but involves vegetation as well and, when doubt exists about where a certain piece of land should be classified, reference is almost always made to the associated vegetation. Nevertheless, since on a broad level of analysis soils and vegetation are correlated closely in Usandawe, the system seems to operate with a fair degree of consistency. It is used to organize the data in this section of the study because of its relevance to understanding conditions as they now exist and also because of the importance of phrasing contemplated changes in terms of what is familiar and understandable to the people. As Allan has put it, if any progress is going to be made in agricultural transformation "we must first try to see the situations through the eyes of the African cultivators. . . ."[20]

The soils traditionally favored by the Sandawe are those associated with *Brachystegia* and *Combretum* woodlands and termed *//'eú*. They are dominantly deep loose-to-friable sands and loamy sands, though a silty texture frequently appears near the base of the hillslopes and where the woodland projects onto level areas. Colors range from shades of gray to browns and reds. The sample profiles given in Table 9 demonstrate a tendency toward fairly high levels of acidity, while organic content is very low, a result attributable in great part to the voracious appetites of termites. Extractable phosphorus is deficient, as are exchangeable bases, excepting potassium. The Sandawe consider these soils to be the best available for bulrush millet, groundnuts, and cassava, while maize, sorghum, and most of the other crops yield adequately in years with good rains. Sweet potatoes are said to be poorly adapted. The real reason for preferring *//'eú* seems to relate more to the ease with which the woodland can be cleared and the soils worked, rather than to any innate fertility advantages. Fire does a good share of the clearing, while the light texture of the soils lend them to being readily pierced with the traditional (but now rarely encountered) digging stick (*tl'weé*) or turned with a hoe (*kolō*).

The term *n//atshá* is used to refer to the soils underlying the upland deciduous thicket communities. Their physical and chemical properties are quite similar to the *//'eú* soils, except for a tendency toward even higher levels of acidity (Table 10). Average horizon pH for the thicket soils is 4.9, compared to 5.3 for the woodland soils. In line with the general similarity between the two soils, the Sandawe draw only minor distinctions between them in terms of the suitability for various crops. Bulrush millet is again the preferred grain, followed by sorghum. Maize apparently does quite poorly. Groundnuts and Bambarra nuts are generally successful and

TABLE 9 Sample Profile Characteristics of //'*Eú* Soils[a]

Depth (in.)[b]	Color	Texture	Structure	Consistency	Permeability	pH	Organic Matter	(ppm)	Exchangeable Bases[c] K	Ca	Mg	Capacity
Site: Songa Hills 1 mi north of Megeyan. Middle hillslope. Dry season.												
0–11	Gray brown	Sandy loam	Weak blocky	Fluffy	Rapid	5.6	1.5	2.3	0.3	1.3	1.6	5.2
11–29	Gray brown	Loamy sand	Moderate blocky	Friable	Rapid	5.3	1.1	1.2	0.2	0.8	1.4	
29–44	Pale brown	Sandy loam	Moderate blocky	Friable	Rapid	5.6	1.0	1.0	0.2	0.8	1.3	
44–55	Pale gray brown	Loamy sand	Moderate blocky	Friable	Rapid	5.8						
55–74	Very pale gray brown	Loamy sand	Weak blocky	Loose	Rapid	6.3			0.1	1.1	1.0	
Site: Sandawe Hills at Bugenika. Middle hillslope. Rainy season, soil damp to 42 in.												
0–9	Very dark brown	Medium sand	Structureless	Loose	Very rapid	5.6	1.5	1.2	0.3	0.7	0.7	1.9
9–24	Medium brown	Sand	Structureless	Loose	Very rapid	5.4	0.6	0.8	0.2	0.4	0.3	0.8
24–42	Medium brown	Sand	Structureless	Loose	Very rapid	5.1	0.7	1.0	0.3	0.3	0.3	0.9
42–59	Light brown	Sand	Weak blocky	Loose	Very rapid	4.8	0.5	0.6	0.2	0.1	0.2	0.6
59–73	Pale gray brown	Sand	Weak crumb	Loose	Very rapid	4.8	0.5	1.0	0.2	0.1	0.3	0.6
Site: Sandawe Hills at Mengu. Lower hillslope. Dry season.												
0–10	Gray brown	Loamy sand	Weak fine grain	Fluffy	Rapid	5.1	0.7	1.5	0.2	0.9	0.1	3.2
10–20	Brownish gray	Loamy sand	Weak fine grain	Fluffy	Rapid	5.3	0.5	1.1	0.2	0.9	0.2	

20–31	Brownish gray	Loamy sand	Weak fine grain	Fluffy	Rapid	5.4	0.2	0.8	0.2	0.6	0.1	
31–48		Loamy sand	Weak fine grain	Fluffy	Rapid	5.5						
48–65	Medium gray	Loamy sand	Structureless	Fluffy	Rapid	5.4			0.1	0.6	0.1	
Site: Sandawe Hills at Samari. Middle hillslope. Dry season.												
0–11	Medium gray	Sand	Weak crumb	Fluffy	Very rapid	5.8	1.3	1.6	0.2	1.3	0.2	3.4
11–20	Pale brown	Sand	Weak crumb	Fluffy	Very rapid	5.4	0.9	0.7	0.1	0.8	0.1	
20–42	Light	Sand	Weak crumb	Fluffy	Very rapid	5.2	0.5	0.4	0.1	0.8	0.1	
42–54	Pale brown	Sand	Weak crumb	Fluffy	Very rapid	5.2						
54–70	Very pale brown	Sand	Weak crumb	Fluffy	Very rapid	5.2			0.1	1.1	0.1	
Site: Sandawe Hills near Ovada. Lower hillslope. Dry season.												
0–6	Gray brown	Silt	Weak blocky	Friable	Rapid	6.6	1.3	1.8	0.4	3.3	0.8	5.6
6–13	Dusky red brown	Silt	Weak blocky	Friable	Rapid	6.1	0.9	1.0	0.4	2.2	1.0	
13–23	Red brown	Silt	Weak grain	Loose	Rapid	5.9	0.7	0.6	0.3	2.1	1.5	
23–42	Rust red	Silt	Structureless	Loose	Rapid	5.8			0.4	1.6	2.8	
42–63	Rust red	Silt	Weak grain	Loose	Rapid	5.7						
63–70	Yellow red	Gravelly sand	Structureless	Loose	Very rapid	5.8			0.2	2.6	1.3	

TABLE 9 *Continued*

Depth (in.)	Color	Texture	Structure	Consistency	Permeability	pH	Organic Matter	(ppm)	Exchangeable Bases K	Ca	Mg	Capacity
Site: Sandawe Hills 3 mi south of Kologa. Level. Rainy season, soil damp to 46 in.												
0–8	Brown	Silt loam	Weak blocky	Friable	Rapid	3.9	1.4	2.2	0.6	0.5	0.6	5.2
8–28	Red brown	Silt loam	Weak blocky	Friable	Rapid	3.6	0.8	2.0	0.5	0.2	0.4	5.6
28–36	Pale red brown	Silt loam	Weak blocky	Friable	Rapid	3.6	0.8	1.4	0.4	0.2	0.3	7.0
36–46	Dusky red	Silt loam	Weak blocky	Friable	Rapid	3.5	0.6	1.6	0.4	0.1	0.3	7.2
46–67	Light gray	Silt loam	Weak blocky	Friable	Rapid	3.4	0.6	1.2	0.3	0.3	0.4	9.3
Site: Sandawe Hills at Farkwa. Middle hillslope. Dry season. Regrowth of 15–20 years.												
0–10	Dusky red	Loamy sand	Weak blocky	Friable	Rapid	5.4	0.9	1.2	0.1	1.2	0.2	4.7
10–19	Dusky red	Loamy sand	Weak blocky	Friable	Rapid	5.3	0.3	0.8	0.1	1.4	0.4	
19–37	Dusky red	Loamy sand	Weak blocky	Friable	Rapid	5.2	0.3	1.3	0.1	0.9	0.9	
37–64	Dusky red	Loamy sand	Weak blocky	Friable	Rapid	5.5			0.1	1.9	1.3	

[a]Physical determinations for these and the soils to follow, except where indicated, were made by the author. Some of the profiles were chemically analyzed at the Northern Research Centre, Tengeru, Tanzania, while the remainder were done by the Soils Testing Laboratory, University of Minnesota. These can be distinguished by the latter's having given complete values for each horizon.

[b]An attempt was made to visually distinguish horizons, thus accounting for variations between profiles in depth.

[c]Million equivalents per 100 grams.

TABLE 10 Sample Profile Characteristics of *N//Atshá* Soils

Depth (in.)	Color	Texture	Structure	Consistency	Permeability	pH	Organic Matter	(ppm)	Exchangeable Bases K	Ca	Mg	Capacity
Site: Sandawe Hills at Sankwaleto. Upper middle hillslope. Rainy season, profile damp to 14 in.												
0–14	Brownish orange	Sand	Structureless	Loose	Very rapid	4.3	1.4	1.7	0.4	0.3	0.4	2.1
14–26	Very dusky orange	Sand	Weak blocky	Friable	Very rapid	4.2	0.9	1.3	0.3	0.3	0.2	2.3
26–49	Medium orange	Sand	Weak blocky	Friable	Very rapid	4.2	0.7	1.2	0.3	0.3	0.3	2.7
49–70	Dusky orange	Sand	Weak blocky	Friable	Very rapid	4.3	0.4	1.6	0.4	0.3	0.2	2.0
Site: Sandawe Hills northeast of Farkwa. Upper middle hillslope. Rainy season, profile damp to 12 in.												
0–6	Brown	Loamy sand	Structureless	Loose	Rapid	4.9	1.4	1.4	0.4	0.2	0.3	1.5
6–12	Red brown	Loamy sand	Structureless	Loose	Rapid	4.5	1.0	2.4	0.6	0.2	0.4	1.5
12–31	Dusky brown	Loamy sand	Weak blocky	Friable	Rapid	4.3	0.7	2.0	0.5	0.1	0.3	2.3
31–51	Medium orange	Loamy sand	Weak blocky	Friable	Rapid	4.5	0.5	3.0+	0.8	0.1	0.1	2.6
51–69	Medium orange	Loamy sand	Weak blocky	Friable	Rapid	4.3	0.5	3.0+	0.8	0.1	0.2	2.4

TABLE 10 *Continued*

Depth (in.)	Color	Texture	Structure	Consistency	Permeability	pH	Organic Matter	(ppm)	Exchangeable Bases K	Ca	Mg	Capacity
Site: Sandawe Hills near Mangasta. Lower middle hillslope. Dry season.												
0–7	Very dusky orange	Sand	Weak blocky	Loose	Very rapid	5.0	1.1	1.6	0.1	0.8	0.1	3.1
7–24	Bright orange	Sand	Moderate blocky	Loose	Very rapid	4.9	0.8	1.4	0.1	0.6	nil	
24–43	Dusky orange	Sand	Moderate blocky	Loose	Very rapid	4.8	0.7	1.3	0.1	0.6	nil	
43–63	Very dusky orange	Sand	Weak blocky	Loose	Very rapid	4.9						
63–80	Very dusky orange	Sand	Weak blocky	Loose	Very rapid	4.9			0.1	0.08	nil	
Site: Sandawe Hills near the road to the north of the Pupuxa. Lower hillslope. Dry season.												
0–14	Very dusky red	Silt loam	Moderate blocky	Fluffy	Rapid	5.4	0.7	1.5	0.4	1.9	0.5	5.7
14–26	Dusky red	Silt loam	Moderate blocky	Fluffy	Rapid	5.1	0.6	0.9	0.4	3.1	0.9	
26–34	Dusky red	Silt loam	Moderate blocky	Fluffy	Rapid	5.1	0.5	0.5	0.2	3.6	1.3	
34–42	Dusky red	Silt loam	Moderate blocky	Friable	Rapid	5.2			0.2	4.8	1.8	
42–56	Dusky red	Silt loam	Moderate blocky	Friable	Rapid	5.6						
56–77	Dusky red	Silt loam	Moderate blocky	Friable	Rapid	7.4			0.3	9.2	2.7	

sweet potatoes are said to grow somewhat better than on *//'eú*, though not appreciably so. Cassava, according to the Sandawe, should not be grown on *n//atshá*. Though *n//atshá* is also light textured and easily worked, the clearing of thicket involves more effort than clearing woodland and consequently, given a choice between the two, the Sandawe will inevitably choose to work the woodland.

Closely allied to *n//atshá* is *n//ansá,* a term applied to the Itigi thicket country. Because of their peripheral and sporadic settlement in this area, the Sandawe have little knowledge of the associated soil conditions, and many individuals do not bother making a distinction between *n//ansá* and *n//atshá*. If pressed about comparative productivity they will simply give the same reply for each or else say they have no idea. The two sample profiles taken—one from the Itigi thicket proper and the other from an outlier near Zarmo—do actually reveal many similar characteristics to the soils under upland deciduous thicket communities (Table 11). The totals for exchangeable bases and extractable phosphorus are extremely low. Acidity is even more pronounced, and while organic matter appears to be slightly more abundant, it is still low. Structure again is poorly developed. However, there are a couple of significant differences in physical makeup. The most striking is the appearance of a silt to silt-loam texture. Also, there is a stronger yellow to red color component that becomes increasingly apparent with depth. Neither profile extended deep enough to encounter the silcrete cement parent material nor the murram layer that reportedly forms just above it.[21]

Throughout the Sandawe Hills there are sporadic occurrences of a soil known to the Sandawe as *kolós'*. Its distinguishing characteristics are a orange-to-rust color and a dominantly silt texture, though in terms of the remaining physical properties and its chemical composition there is nothing that separates it from the other upland soils (Table 12). Nevertheless, the Sandawe accord *kolós'* with quite different capabilities. Sorghum is said to grow very well, while both bulrush millet and maize supposedly do poorly. Groundnuts and Bambarra nuts produce satisfactorily but not cassava. The real forte of *kolós',* though, is its ability to support sweet potatoes, a fact evidently related to the greater moisture retention capabilities of silts over sands. Because of this advantage, it is extremely difficult to find an area of *kolós'* that has not been extensively cut over and cultivated. Much now seems to have reached a state of deterioration, as indicated by widespread gullying and patches of *Dichrostachys cinerea* thicket.

Gáwa is the Sandawe name for hill, but it is also used to denote the soils found on steep hillsides and in depressions along the ridge crests, in other

TABLE 11 Sample Profile Characteristics of *N//Ansá* Soils

Depth (in.)	Color	Texture	Structure	Consistency	Permeability	pH	Organic Matter	(ppm)	Exchangeable Bases K	Ca	Mg	Capacity
Site: Near Ageba. Level. Rainy season, profile wet to 6 in. and damp to 27 in.												
0–6	Medium brown	Silt loam	Structureless	Loose	Rapid	4.8	2.5	2.1	0.5	1.3	0.6	4.4
6–15	Very dusky orange	Silt loam	Weak crumb	Fluffy	Rapid	4.0	1.3	1.9	0.5	0.3	0.3	3.7
15–27	Dusky orange	Silt loam	Weak blocky	Fluffy	Rapid	3.8	1.0	1.8	0.4	0.2	0.3	4.5
27–44	Dusky orange	Silt loam	Weak blocky	Fluffy	Rapid	3.7	0.7	2.0	0.5	0.2	0.3	5.0
44–69	Very dusky orange	Silt	Structureless	Loose	Rapid	3.6	0.9	2.6	0.7	0.6	0.6	5.2
Site: Between Sanzawa and Mbuyuni. Level. Dry season.												
0–11	Light brown gray	Silt	Weak crumb	Friable	Rapid	4.8	1.2	1.5	0.2	0.9	0.3	4.4
11–22	Pale yellow brown	Silt	Weak crumb	Friable	Rapid	4.8	1.0	0.9	0.2	0.6	0.2	
	Yellow brown	Silt	Weak crumb	Fluffy	Rapid	4.8	0.3	0.5	0.2	0.6	0.1	
33–45	Yellow brown	Silt	Weak crumb	Fluffy	Rapid	4.8			0.1	0.7	0.1	
45–69	Light red brown	Silt	Weak crumb	Fluffy	Rapid	4.8			0.2	0.6	0.4	

TABLE 12 Sample Profile Characteristics of *Kolós'* Soils

Depth (in.)	Color	Texture	Structure	Consistency	Permeability	pH	Organic Matter	(ppm)	Exchangeable Bases K	Ca	Mg	Capacity
Site: Sandawe Hills at Doyo. Lower hillslope. Dry season.												
0–14	Dusky red orange	Silt	Weak crumb	Friable	Rapid	5.0	0.9	1.6	0.5	0.7	0.4	1.8
14–26	Rust red	Silt	Weak crumb	Friable	Rapid	4.6	0.4	1.1	0.4	1.0	0.4	
26–39	Rust red	Silt	Weak crumb	Friable	Rapid	4.5	0.4	0.6	0.4	1.0	0.2	
39–51	Rust red	Silt	Weak crumb	Friable	Rapid	4.6						
51–63	Rust red	Silt	Weak crumb	Friable	Rapid	4.7			0.2	1.5	0.6	
Site: Sandawe Hills near Sankwaleto. Middle hillslope. Rainy season, profile damp to 31 in.												
0–8	Red brown	Silt loam	Structureless	Loose	Rapid	4.6	1.6	2.7	0.7	0.6	0.9	3.3
8–19	Red brown	Silt loam	Structureless	Loose	Rapid	4.0	1.0	2.0	0.6	0.4	0.6	5.4
19–32	Red orange	Silt loam	Weak blocky	Fluffy	Rapid	4.0	0.8	1.0	0.2	0.3	0.7	6.8
32–54	Bright orange	Silt loam	Moderate blocky	Friable	Rapid	3.8	0.7	1.1	0.3	0.6	1.1	5.7
54–69	Bright orange	Sandy silt loam	Moderate blocky	Friable	Rapid	4.0	0.6	1.5	0.4	0.3	1.2	6.5

words, those associated with the Ridge Crest complex of vegetation. These thin, quite stony soils are of a well-developed loam texture at the surface, grading into a somewhat sticky sandy clay loam just above the decaying bedrock (Table 13). Surface organic matter is a little higher than among other upland soils and so is pH. The sample profile taken from a hilltop depression near Kologa has the highest readings of extractable phosphorus of any soil so far recorded in Usandawe. The crops grown in *gáwa* include all three main grains plus some eleusine, and the Sandawe universally claim that they all grow better here than anywhere else. These are considered the cream of Sandawe soils. Necessarily, the fields are restricted in size, ranging from as low as a few square yards up to an acre or more in some of the large depressions.

The mainly gray-brown soils found in conjunction with *Acacia-Commiphora* Woodland and Thicket are always given the name *xáts'a.* Texturally, there is considerable variation between profiles and within a given profile, but in comparison with all the soils discussed until now, texture is noticeably heavier at every level (Table 14). Clay becomes particularly dominant in the subsoil, forming the hardpan layer at a depth of about 2 ft. Though often waterlogged and sticky above the hardpan during the rains, *xáts'a* soils bake to a virtual cement that is almost impossible to break during the dry season. The available bases and potential capacity show an improvement over the lighter textured soils, as does pH, which assumes a more neutral value. Organic matter and phosphorus are once more very low.

Closer inspection of the sample profiles indicates that there are at least two types of *xáts'a,* depending on whether the site is located in the interhill drainage lines and depressions or on the plains surrounding the hills. The latter tend to have heavier textures, more basic pH's, a somewhat greater percentage of organic matter, and a more fully developed quantity of exchangeable bases. It should be expected, of course, that the interhill *xáts'a* would more closely resemble the upland soils, since each year they receive a fresh supply of rainwash from above.

The Sandawe traditionally have not utilized the *xáts'a* soils to any extent for cultivation, mainly because of the greater amount of effort that must be expended on them in comparison to the lighter soils. Planting has to await the onset of the rains so that the topsoil becomes loose enough to work with a hoe, and cultivation is rendered considerably more difficult in a soil of heavy texture and sticky consistence. Furthermore, clearing is complicated by the dense stands of thorn thicket. The only major exception to this rule is the Dantl'awa area where *xáts'a* has been cropped for many years. But as the sample profile from there indicates, texture approaches that for the

TABLE 13 Sample Profile Characteristics of *Gáwa* Soils

Depth (in.)	Color	Texture	Structure	Consistency	Permeability	pH	Organic Matter	(ppm)	Exchangeable Bases K	Ca	Mg	Capacity
Site: Sandawe Hills near Farkwa. Upper hillslope. Rainy season, profile damp to 5 in.												
0–5	Medium brown	Loam	Weak blocky	Friable	Moderate	6.2	2.7	1.9	0.5	3.3	1.2	4.7
5–25	Dusky brown	Sandy loam	Moderate blocky	Friable	Rapid	5.4	1.4	0.5	0.1	2.1	0.7	3.3
25–45	Rust red	Sandy clay loam	Moderate blocky	Friable	Moderate	5.6	0.9	0.5	0.1	1.2	1.5	4.2
45–60	Rust red	Sandy loam	Weak blocky	Friable	Rapid	6.2	0.9	0.7	0.1	3.3	0.8	3.9
60–75	Orange	Gravel—decaying bedrock				7.1	0.9	1.1	0.3	5.2	1.2	
Site: Sandawe Hills at Kologa. Hilltop depression. Dry period of several weeks during rains, profile dry.												
0–8	Very dusky brown	Loam	Moderate blocky	Friable	Moderate	5.7	2.1	2.3	0.6	2.4	1.1	5.0
8–19	Dusky brown	Loam	Moderate blocky	Friable	Moderate	5.3	1.2	2.1	0.5	1.2	1.0	5.0
19–28	Rust red	Sandy clay loam	Moderate blocky	Friable	Moderate	4.7	0.9	3.0	0.8	1.2	1.0	5.5
28–64	Rust red	Gravel—decaying bedrock				4.7	0.6	2.8	0.7	1.8	0.8	4.7

TABLE 14 Sample Profile Characteristics of *Xáts'a* Soils[a]

Depth (in.)	Color	Texture	Structure	Consistency	Permeability	pH	Organic Matter	(ppm)	Exchangeable Bases K	Ca	Mg	Capacity
Site: Sandawe Hills at Dantl'awa. Level. Dry season.												
0–7	Pale gray brown	Loamy sand	Moderate blocky	Friable	Moderate	5.7	0.7	1.3	0.3	1.2	0.6	6.4
7–19	Pale gray brown	Loamy sand	Moderate blocky	Friable	Moderate	5.2	0.4	0.4	0.2	1.8	0.8	
19–30	Pale brown	Silt loam	Strong blocky	Hard	Slow	5.2	0.7	0.4	0.2	2.8	1.4	
30–36	Pale brown	Silt loam	Strong blocky	Hard	Slow	5.3			0.2	2.8	1.5	
36–41	Very pale gray	Silt loam	Strong blocky	Hard	Slow	5.8						
41–51	Pale gray brown	Silt loam	Strong blocky	Hard	Slow	8.3			0.2	10.6	3.3	
Site: Sandawe Hills near Tumbakose. Level. Rainy season, profile wet to 16 in. and damp to 27 in.												
0–9	Gray brown	Sandy clay loam	Weak blocky	Loose	Moderate	4.4	1.5	2.0	0.4	2.1	2.4	5.5
9–16	Gray brown	Sandy clay loam	Weak blocky	Loose	Moderate	4.4	0.8	0.9	0.2	1.6	1.8	4.5
16–27	Medium brown	Medium clay	Moderate blocky	Plastic	Slow	4.7	1.3	1.0	0.2	2.7	2.3	9.6
27–52	Pale gray brown	Sandy clay	Moderate blocky	Firm	Moderate	8.0	0.5	1.5	0.4	9.9	3.5	9.3
52–69	Pale gray	Silty clay loam	Moderate blocky	Friable	Moderate	8.1	0.5		0.3	15.8	4.6	13.7
Site: Just west of Ngaia. Level. Dry season, but profile slightly damp from 35–54 in.												
0–6	Pale gray brown	Sandy clay loam	Moderate blocky	Compact	Moderate	5.8	1.4	1.4	0.2	10.8	5.9	19.3

6–20	Very pale gray brown	Sandy clay loam	Moderate blocky	Hard	Slow	6.8	0.9	1.1	0.2	13.2	6.1	
20–35	Gray brown	Medium clay	Moderate blocky	Hard	Very slow	5.8	0.4	1.8	0.2	14.0	6.2	
35–54	Gray brown	Medium clay	Moderate blocky	Hard	Very slow	7.6						
54–63	Medium gray	Clay loam	Weak grain	Friable	Moderate	8.7			0.2	38.6	7.5	
Site: 3 mi northwest of Babayu. Level. Dry spell of 10 days after initial rains, profile dry.												
0–10	Gray brown	Sandy clay loam	Moderate blocky	Friable	Moderate	5.9	2.7	1.3	0.2	4.1	2.5	6.9
10–26	Medium gray	Medium clay	Moderate blocky	Friable	Slow	8.3	1.2	1.5	0.4	17.8	5.9	15.8
26–45	Medium gray	Medium clay	Moderate blocky	Friable	Slow	8.3	0.8	1.7	0.4	19.8	5.1	16.7
45–61	Medium gray	Medium clay	Moderate blocky	Friable	Slow	8.6	0.8	1.8	0.5	22.5	5.9	20.7
Site: Near Poro. Level. Dry season, but profile damp from 37–50 in.												
0–8	Medium gray	Sandy clay loam	Moderate blocky	Friable	Moderate	6.3	1.3	1.6	0.5	8.1	1.0	9.6
8–13	Gray brown	Sandy clay loam	Moderate blocky	Friable	Moderate	5.6	0.8	1.2	0.2	8.8	3.1	
13–37	Gray brown	Medium clay	Strong blocky	Compact	Slow	6.7	1.0	1.3	0.2	9.5	3.4	
37–50	Gray brown	Medium clay	Strong blocky	Compact	Slow	7.7			0.2	11.8	4.1	

TABLE 14 *Continued*

Depth (in.)	Color	Texture	Structure	Consistency	Permeability	pH	Organic Matter	(ppm)	Exchangeable Bases K	Ca	Mg	Capacity
Site: South of Bangani.[a] Level. Rainy season, profile wet to 9 in.												
0–4	Dark gray brown	Sandy clay		Friable	Moderate	7.4	1.9		0.6	8.0	5.6	17.4
4–9	Dark brown	Sandy clay		Friable	Moderate	7.0						
9–18	Dark gray	Medium clay		Hard	Slow	8.9						
18–37	Gray brown	Medium clay		Hard	Slow	8.0						
37–51	Medium brown	Medium clay		Hard	Slow	9.1						
51–65	Gray brown	Medium clay		Hard	Slow	9.1						
Site: Kioli.[a] Level. Rainy season, profile damp to 11 in.												
0–3	Very dark brown	Clay loam		Friable	Moderate	7.3	3.4		1.0	14.2	3.8	19.4
3–11	Dark gray	Silty clay		Friable	Moderate	7.4	1.4					
11–20	Dark gray	Silty clay		Hard	Slow	9.0	1.7					
20–37	Dark gray	Silty clay		Friable	Moderate	9.1						
37–46	Very dark gray	Silty clay		Friable	Moderate	9.2						
46–53	Medium gray	Silty clay		Friable	Moderate	9.3						

Site: Kangataro.[a] Level. Rainy season, profile damp to 4 in.												
0–4	Dark gray brown	Loamy sand		Soft	Rapid	7.2	1.4		0.3	5.1	2.5	9.3
4–15	Dark gray brown	Loamy sand		Friable	Rapid	9.6	0.9					
15–28	Gray brown	Loamy sand		Friable	Rapid	9.6						
28–43	Light brown	Sandy clay		Friable	Moderate	9.7						
43–63	Light brown	Sandy clay		Friable	Moderate	9.8						
Site: Ndoroboni. Level. Dry season, but profile slightly damp from 34–68 in.												
0–10	Gray	Silt loam	Strong blocky	Hard	Moderate	7.1	1.0	1.4	0.2	9.4	5.1	16.4
10–21	Gray	Silt loam	Strong blocky	Hard	Moderate	8.0	0.7	1.6	0.1	10.8	5.2	
21–34	Gray	Clay loam	Strong blocky	Compact	Slow	8.8	0.7		0.2	14.4	6.4	
34–47	Gray	Clay loam	Strong blocky	Compact	Slow	9.2			0.2	26.4	7.3	
47–58	Gray	Clay loam	Strong blocky	Compact	Slow	9.3						
58–68	Gray	Clay loam	Strong blocky	Compact	Slow	9.1			0.2	46.4	8.1	

[a]Data provided by Mr. John English, Division of Lands and Settlement, Dar es Salaam, Tanzania.

upland soils and thus the ground is more easily worked. Nevertheless, within the last decade or so there has been a move by the Sandawe to bring *xáts'a* soils into their agricultural system. Two basic reasons appear to be behind the change. First, there was the resettlement of families from the tsetse-infested Songa Hills to clearings between Poro and Ndoroboni and near the Bubu-Mkinke junction, a facet of Sandawe history that will be elaborated on shortly. Second, and seemingly more important, has been the increasing food-reliance placed on maize. The Sandawe recognize that maize is more water-demanding than the other grains and that *xáts'a* soils have better water-holding capabilities than the more freely drained sands and loamy sands. Consequently, fields have been springing up in *xáts'a* areas across the Sandawe Hills. Some individuals, particularly around Tumbakose, have completely given up cultivating the hillsides, while others are beginning to plant at least two grain fields, one upslope and one downslope. Sorghum has been found to grow well in *xáts'a* and so it is usually interplanted with maize, but bulrush millet does not thrive and is kept on the lighter textured soils exclusively. Sweet potatoes, groundnuts, and Bambarra nuts have encountered fair success. Cassava appears to do poorly.

It should be stressed at this juncture that there is still a considerable amount of resistance to cultivating heavy soils, and most of the *xáts'a* land remains untouched. There has been no mass movement of Sandawe downslope, but rather a gradual encroachment. Most of the maize expansion has taken place on *//'eú* and *//atshá* soils.

The final major soil type is known as *thõ,* a term which corresponds to soils associated with mbugas or Valley Grasslands. As the sample profiles demonstrate, in Usandawe they vary from neutral to highly basic in reaction, with a frequent occurrence of soluble salts in the subsoil (Table 15). Despite the prevalent dark color, organic matter is low. Because of the high clay content, permeability is extremely slow, making for the well-known impassability of such areas during the rains by any means other than on foot. Vehicles have been known to remain mired until the next dry season. The Sandawe consider *thõ* to be uncultivable.

Though minor in areal extent, mention should also be made of termite-mound soils (Figure 24). These organic-rich outliers in a sea of sterility are coveted by the Sandawe for furnishing a fairly waterproof filler between the log frame of the *tembe* and for cropping, once the mounds have been rounded and smoothed by erosion. Before its recent expansion, maize was restricted to old mounds and burned debris patches in first-year fields.

From an overall agronomic point of view, it would appear safe to conclude that edaphic conditions in Usandawe leave much to be desired. With-

FIGURE 24 Termite mounds are a common sight in Usandawe. Their earth is used for house construction and their sides are cultivated when they become rounded-down with age. This one stands over 8 ft high.

out exception, the soils are deficient in organic matter, extractable phosphorus, and exchangeable bases, except potassium. A considerable investment in fertilizers, particularly of the nitrogenous and phosphate varieties, would be needed to bring them up to profit-making standards. Wheat, thus, becomes unfeasible as a cash crop. The upland soils and those of the Itigi thicket suffer further from high levels of acidity, poorly developed structure, and excessive drainage. The best are the *gáwa* types, but as the leading authority in the study of Tanzanian soils has put it, the areas which they underlie are best "looked upon as regulators of the runoff . . . and should be kept under bush."[22] Though more attention should undoubtedly be given to the potential of *xáts'a* soils, there is another serious drawback other than those already mentioned. The hardpan layer prevents the penetration of moisture to any great depth and, if heavy rains are followed by long dry spells (as so frequently happens in this country), the plants, so to speak, will be left high and dry. There is no moisture reserve that might be tapped. With reference to the *mbuga* soils, the Sandawe seem to have assessed their potential correctly. Because of the severe drainage problem there is little chance of making them productive in the near future. Fortunately, they are not too extensive in Usandawe.

The soil difficulties are clearly great, yet, because of natural differences, there is some crop specialization, and this can probably be further refined and expanded. There is a little flexibility for the agricultural planners to play with.

In conclusion, for future reference purposes it will be useful to place the soils of Usandawe into some recognized classificatory systems. For

TABLE 15 Sample Profile Characteristics of *Thõ* Soils[a]

Depth (in.)	Color	Texture	Structure	Consistency	Permeability	pH	Organic Matter	(ppm)	Exchangeable Bases K	Ca	Mg	Capacity
Site: 1 mi east of Kwa Mtoro. Level. Dry season, but profile damp from 14–69 in.												
0–14	Dark gray	Medium clay	Strong blocky	Intractable	Slow	8.1	1.4	1.3	1.0	42.6	18.5	55
14–25	Black	Heavy clay	Strong blocky	Sticky	Very slow	7.7	0.9	2.2	0.9	66.6	19.0	
25–40	Black	Heavy clay	Strong blocky	Sticky	Very slow	7.7	1.1	2.8	0.9	68.1	22.3	
40–54	Black	Heavy clay	Strong blocky	Sticky	Very slow	7.8						
54–69	Black	Heavy clay	Strong blocky	Sticky	Very slow	7.7			1.0	59.5	24.8	
Site: Kioli.[a] Level. Rainy season, profile damp to 8 in.												
0–8	Black	Medium clay		Friable	Slow	8.4	1.7		1.0	56.2	18.1	47

8–28	Black	Medium clay	Hard	Very slow	9.2		0.5	44.9	15.5	
28–45	Black	Medium clay	Hard	Very slow	9.0					
45–55	Light gray	Sand	Loose	Rapid	9.1		0.7	54.7	9.6	
Site: Near junction of Bubu and Mkinke.[a] Level. Rainy season, profile damp to 5 in.										
0–1	Very dark brown	Sandy clay	Friable	Moderate	7.3	2.6	1.3	16.2	4.5	21.2
1–5	Black	Medium clay	Hard	Slow	7.5		1.0	21.6	4.8	
5–12	Black	Heavy clay	Hard	Very slow	8.2		0.8	29.7	4.1	
12–25	Black	Heavy clay	Very hard	Very slow	7.6					
25–36	Black	Heavy clay	Very hard	Very slow	7.7					
36–56	Black	Heavy clay	Very hard	Very slow	7.4					
56–72	Black	Heavy clay	Very hard	Very slow	7.7		1.1	31.9	5.2	

[a]Information from Mr. John English.

Africa as a whole, the most recent and now most generally accepted system is the one constructed by the CCTA symposium held at Leopoldville in 1963.[23] According to this scheme, all the uplands soils—*//'eú, n//atshá, kolós',* and *gáwa*—plus those of the Itigi thicket (*n//ansá*) would be classed as Ferrisols or perhaps Ferralitic for the very acid ones. *Xáts'a* fits more closely in the Calcimorphic soils category and *thõ* would be considered a Vertisol. For Tanzania, a system has been devised by Anderson.[24] He would place *//'eú, n//atshá, kolós',* and *gáwa* under the heading of Non-laterized Red and Yellowish soils. The first three fit the subheading of soils developed on metamorphic rocks and granites and the latter those developed on grits. *N//ansá* gets special recognition as a Palaeosol. The closest approximation to *xáts'a* is the category Brown Soils of High Base Status, though the base statuses of those in Usandawe do not reach the levels of Anderson's examples. Lastly, *thõ* would be put with the Tropical Black Earths of Calcareous Nature. Map 7 presents a breakdown slightly modified from Anderson.

DISEASES AND PESTS

An extremely important consideration in crop production concerns the various diseases and pests that prey upon both man and plants. For man, they are most noticeable in affecting his ability to work efficiently and effectively, while on plants they usually act to reduce yields, though on occasion almost complete destruction can take place. It is beyond the scope of the present study to go into the ecology, prevention, and cure of the diseases and pests menacing Usandawe, but an indication of at least the most serious occurrences would seem desirable.

Endemic malaria is unquestionably the most serious human disease problem.[25] It is the primary killer, particularly of children, and brings periodic disability to an estimated one-half of the population. The peak season is during the latter portion of the rains when numerous pools of standing water offer ideal breeding for the *Anopheles* mosquito. Unfortunately, this is the same time of year when guarding the fields requires the most effort and when labor will shortly be needed for the harvest. The Sandawe themselves admit that malaria contracted at this time does result in serious hardship for individual families. Tuberculosis is very common among the middle-aged and elderly segments of the population. Its debilitating effects seriously reduce the infected individual's capacity for work. Relapsing or tick fever is widespread, though it is not as fatal as it is to Europeans. Along with most Africans, the Sandawe have apparently built up a certain amount of immunity. The ticks burrow in the dirt floors of the *tembe* and, if sleeping mats instead of beds are used, they can become a serious hazard. As the name implies, the fever recurs periodically incapacitating the victim for several days or weeks. Bilharzia has been spreading rapidly across central Tanzania in recent years, and it is now probably safe to assume that all standing water is infected. The snail,

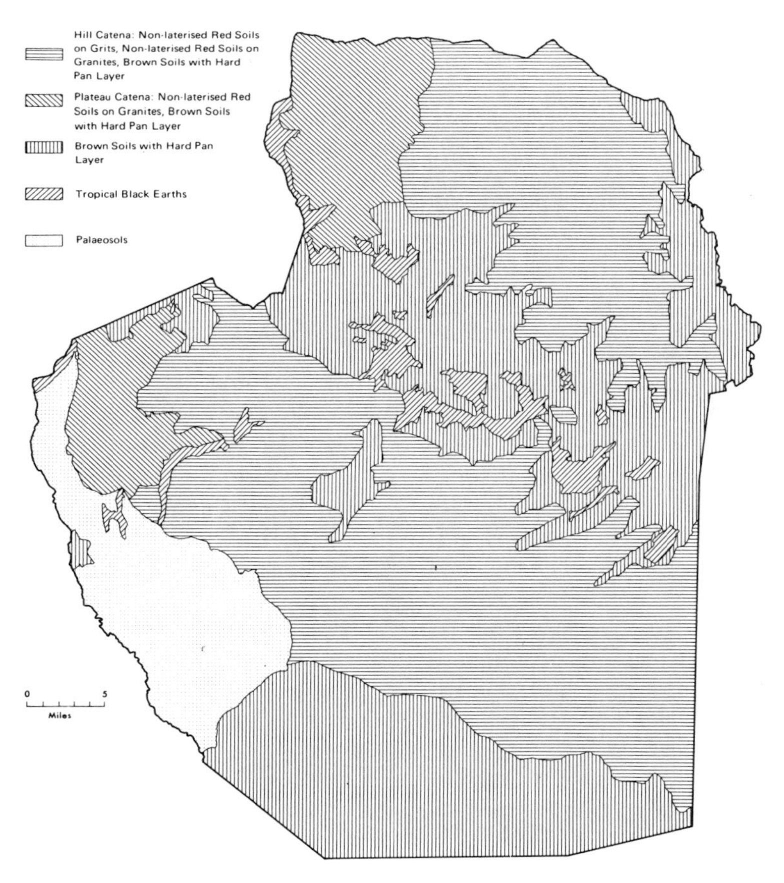

MAP 7 Soil types of Usandawe.

which acts as the intermediate host, combines the ability of surviving long dry periods with that of being able to travel rather great distances to find suitable water. Various forms of dysentery are chronic and periodically disable the majority of people, while some amoebic varieties cause a very rapid and violent death. Polio and meningitis are present, though few deformities are seen. Evidently these diseases are fatal in most instances. Whooping cough and chicken pox occasionally result in death for young children.

Many of the more virulent diseases of Africa—yellow fever, typhus, diphtheria, typhoid, rabies, and tetanus—are virtually unknown. Small pox formerly was very prevalent, but now seems to be under control. Sleeping sickness of the variety *Trypanosomiasis rhodesiense* has not occurred since the 1940's.

Of the less immediately fatal or debilitating types, venereal diseases are the most common. Both syphilis and gonorrhea are endemic and are undoubtedly responsible for a good share of the widespread blindness and feeblemindedness. Conjunctivitis, a staphylococcus infection of the eye, constitutes the other major source of blindness. Elephantiasis is not common. In a year's time I saw only one case, and it was a mild one causing no noticeable impairment. Occasionally, leprosy is reported, though for some reason the Sandawe do not appear to be as susceptible as many other peoples in central Tanzania. The Turu, for example, are reported to have an incidence of about 2 percent.[26] Hookworm is a constant minor problem, while roundworm and tapeworm are very sporadic in occurrence. Malnutrition is not generally serious, except, obviously, during the periodic famines. In most years the quantity of food is apparently adequate. Also, there are some indications that quality is not too bad. The distended belly of kwashiorkor is rarely seen, evidencing that protein intake is adequate. Vitamin C deficiency and pellagra, both common in central Tanzania, are again infrequent. Perhaps the intake of bush fruits plays a significant role in holding down their incidence.

Birds, the primary living menace to crops, have already been discussed. However, there are several other pests that merit some consideration. Among the insects, the desert locust constitutes the main threat for possible large scale destruction. It has not come south of Arusha in any numbers since the early 1950's, and actually is reported to be in a state of retreat.[27] But we may be witnessing only a temporary phase, and it is probably best to consider the ravenous creature as a constant potential invader. In contrast, red locust and migratory locust give all indications of being under control.[28] Maize is susceptible to attacks and consequent yield reduction by the stalk borer *Busseola fusca.*[29] Large numbers survive each year because the stubble is left in the fields or sometimes used in house construction. The green shield bug (*Nezara viridula*) similarly infests sorghum and bulrush millet. It proliferates when rainfall reaches above 30 in. and produces two generations in one growing season. Similarly, rosette in groundnuts, which is spread by *Aphis craccivora,* is most serious during exceptionally wet years. Increased rainfall is not always an unmixed blessing. Mosaic in castor is caused by a white fly of the *Bemisia* genus. It drastically reduces yields. Castor has two destructive enemies. *Lygus vasseleri* is a sucking bug that attacks the flowering spike. The established giant castor varieties are somewhat resistant, but introduced dwarf variants were devastated and consequently are no longer grown. *Helopeltis schoutedeni* from time to time appear in enormous numbers, causing damage to the spike and retarding growth in general. At present, cotton is virtually precluded by the presence of sucking bugs belonging to the genera *Calidea* and *Dysdercus.*

Some leaf rust occurs on maize and smut is occasionally seen on sorghum and bulrush millet, though neither constitutes a serious problem. During exceptionally wet years the latter two grains develop a fungus growth known in East Africa as *asali* disease. The term is Swahili for honey and refers to the sweating out of sugar by the plants. Fairly substantial yield reductions can be expected when there is an outbreak.

Weevils yearly take their share of stored grains, particularly maize. After threshing, all grains are placed in large circular bins made from the bark of *Brachystegia spiciformis.* Despite the fact that the bins are raised several inches off the ground on planks, they are not particularly successful in excluding weevils, nor for that matter small rodents. It is impossible to estimate the loss, though we can say that each family experiences some reduction in total food supply.

Of the larger animals, six can be singled out for the havoc they create. Certainly far and away the greatest four-footed pest is the elephant. A migrating herd can easily trample several fields in one night's escapade, in addition to which they have a fondness for snatching pumpkins and rooting up sweet potatoes. Without guns, the Sandawe have virtually no defense against them. Some people try burning a patch around their fields to discourage foraging, but the more usual method is to make as much noise as possible when a herd is known to be nearby. Baboons and monkeys rank next in the hierarchy. They show a well-developed taste for maize, sweet potatoes, groundnuts, cassava, and pumpkins. A similar diet is demonstrated by bush pigs, porcupines, and spring hares. Noise is once more applied as the major defensive measure against these intruders, along with trapping and hunting. Thorn-branch barriers are frequently circled around fields, but have not proved much of an obstacle. The spring hare and porcupine add a further dimension to the control problem by concentrating their activities at night. A more complete list of animal pests is given in Table 16.

TABLE 16 Animal Pests of Usandawe

Common Name	Scientific Name	Sandawe Name	Damage Inflicted
Elephant	*Loxodonta africana*	*n/uaá*	Tramples fields. Eats maize, melons, sweet potatoes
Yellow baboon	*Papio cynocephalus*	*//o//á*	Eats all grains, melons, sweet potatoes, and cassava
Vervet monkey	*Ceropithecus aethiops*	*!haáku*	Eats all grains, melons, sweet potatoes, and cassava
Bush pig	*Potamochoerus choeropatamus*	*kéuto*	Eats sweet potatoes, cassava, groundnuts, and Bambarra nuts
Porcupine	*Hystrix galeata*	*kialéé*	Eats melons, sweet potatoes, cassava, groundnuts, and Bambarra nuts
Spring hare	*Pedetes surdaster*	*di'ra*	Eats cassava, sweet potatoes, groundnuts, and Bambarra nuts
Greater kudu	*Tragelaphus strepsiceros*	*g!okomi*	Eats castor flowers
Bat-eared fox	*Otocyon megalotis*	*búguli*	Eats groundnuts and Bambarra nuts
Genet	*Genetta genetta*	*thandeé*	Eats groundnuts and Bambarra nuts
Dwarf mongoose	*Helogale undulata*	*tsuaámaa*	Occasionally eats groundnuts and Bambarra nuts
White-tailed mongoose	*Ichneumia albicauda*	*nima*	Eats chickens and occasionally groundnuts and Bambarra nuts

TABLE 16 *Continued*

Common Name	Scientific Name	Sandawe Name	Damage Inflicted
Banded mongoose	*Mungos mungo*	*xare*	Eats chickens and occasionally groundnuts and Bambarra nuts
Civet cat	*Civettictis civetta*	*zorí*	Eats chickens and occasionally groundnuts and Bambarra nuts
Black-backed jackal	*Canis mesomelas*	*monzó*	Eats chickens and occasionally groundnuts and Bambarra nuts
Serval cat	*Felis serval*	*ababúwaa*	Eats chickens
African wildcat	*Felis lybica*	*káxá*	Eats chickens
Black-tipped mongoose	*Herpestes sanguineus*	*lóngos'*	Eats chickens
Cafer cat	*Felis ocreata*	*páka*	Eats chickens
Honey badger	*Mellivora capensis*	*zírba*	Eats chickens and robs beehives
Spotted hyena	*Crocuta crocuta*	*thékele*	Eats chickens and the young of cattle, sheep, and goats and occasionally maize
Lion	*Panthera leo*	*//atsú*	Eats cattle, sheep, goats, and donkeys
Leopard	*Panthera pardus*	*theká*	Eats dogs, goats, sheep, donkeys, and young cattle
Cheetah	*Acinonyx jubatus*	*babaaro*	Eats dogs, goats, and sheep
Four-striped grass mouse	*Rhabdomys pumilio*	*molóto*	Eats stored grains
Spring mouse	*Acomys selousi*	*méba'e*	Eats stored grains
Unstriped grass mouse.	*Arvicanthis abyssinicus*	*zakha'búr'*	Eats stored grains

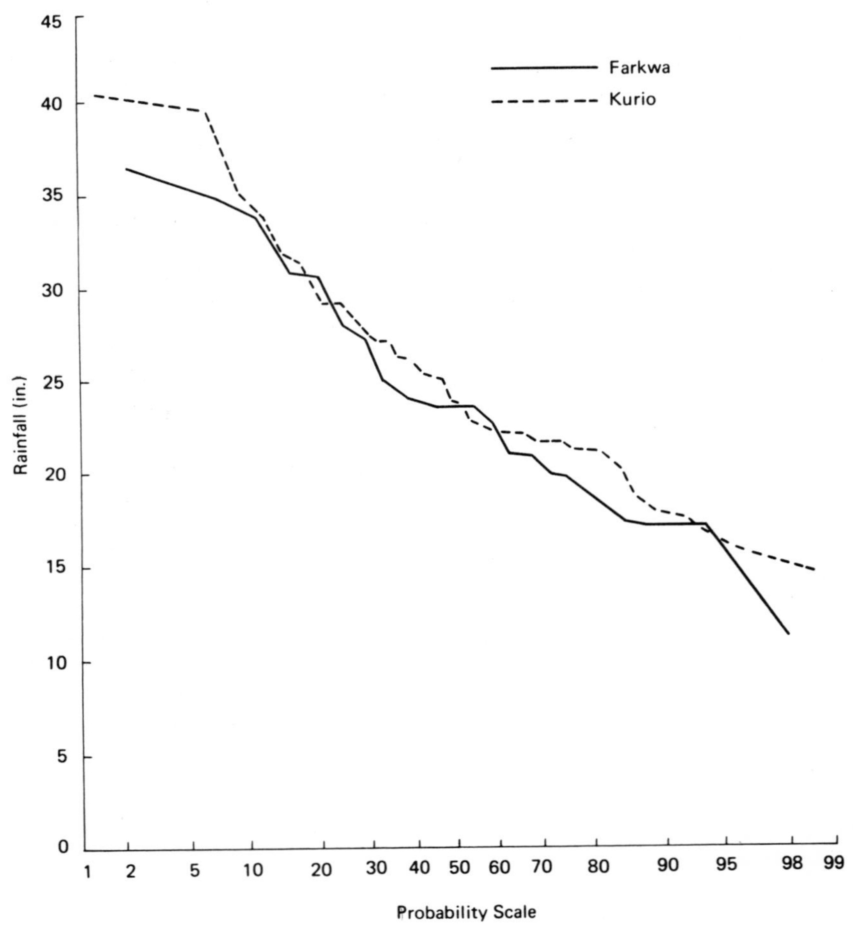

GRAPH 1 Annual rainfall probability.

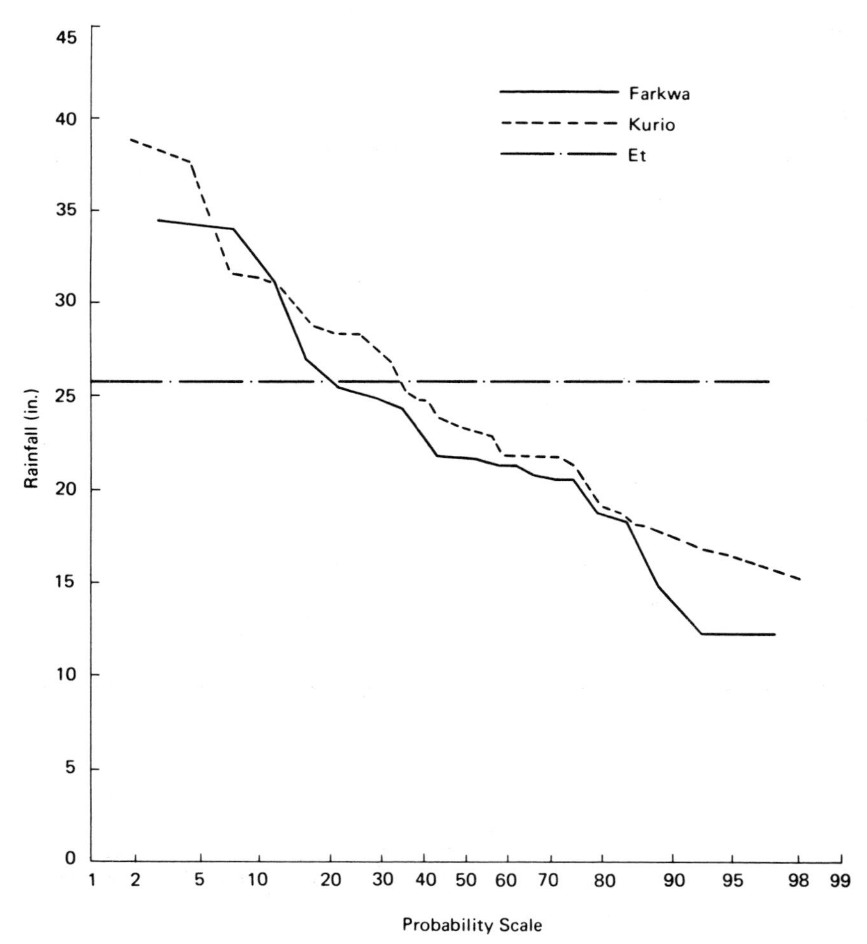

GRAPH 2 Growing season rainfall probability.

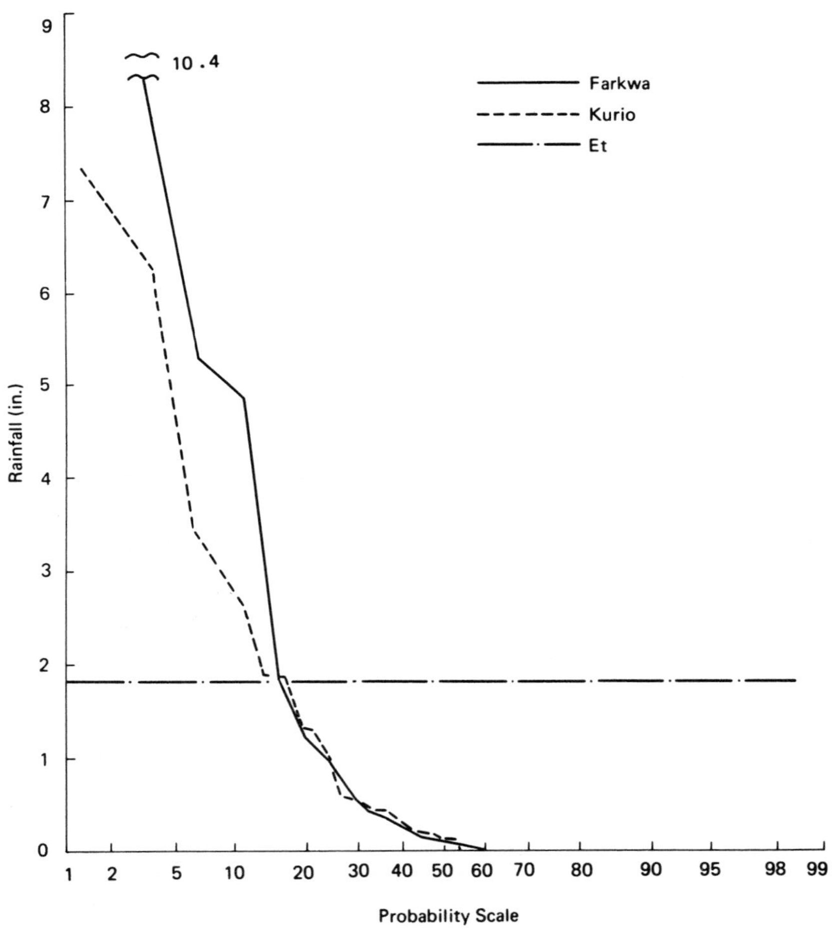

GRAPH 3 Rainfall probability for November III.

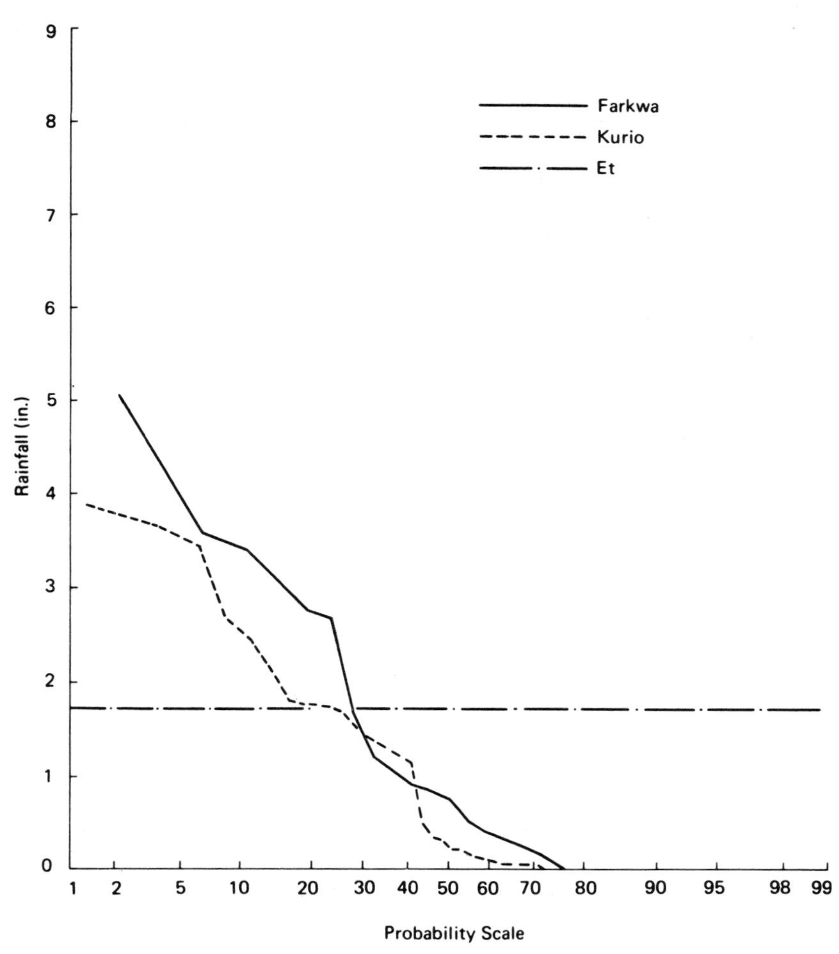

GRAPH 4 Rainfall probability for December I.

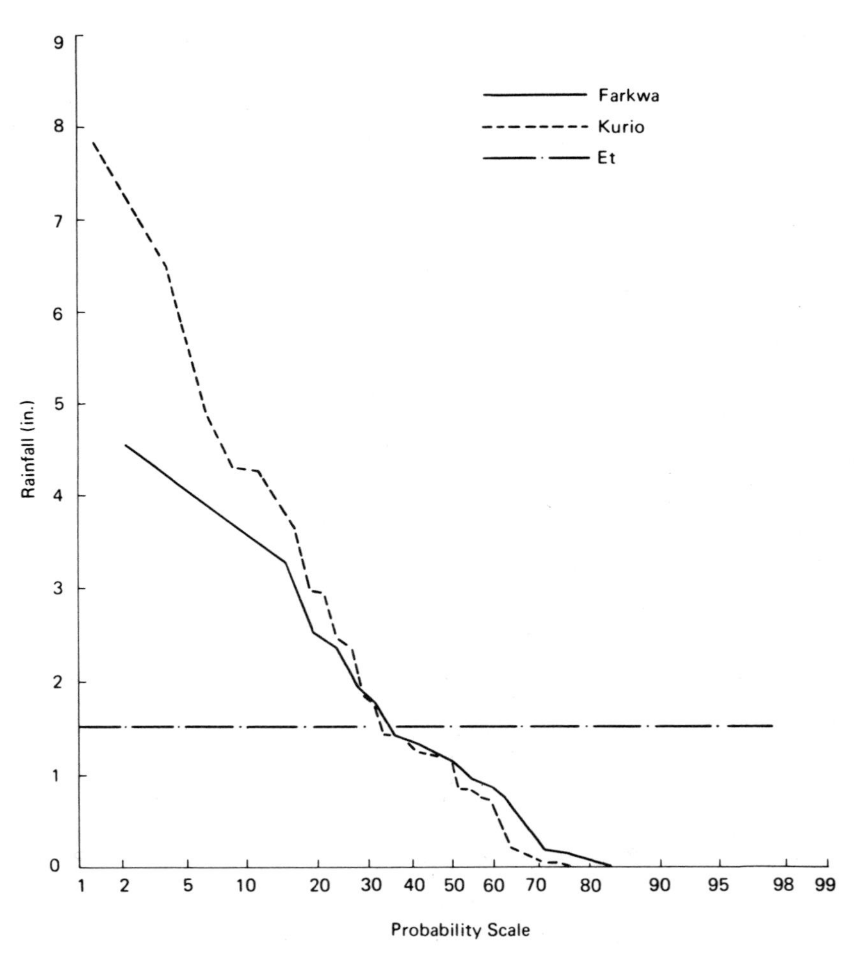

GRAPH 5 Rainfall probability for December II.

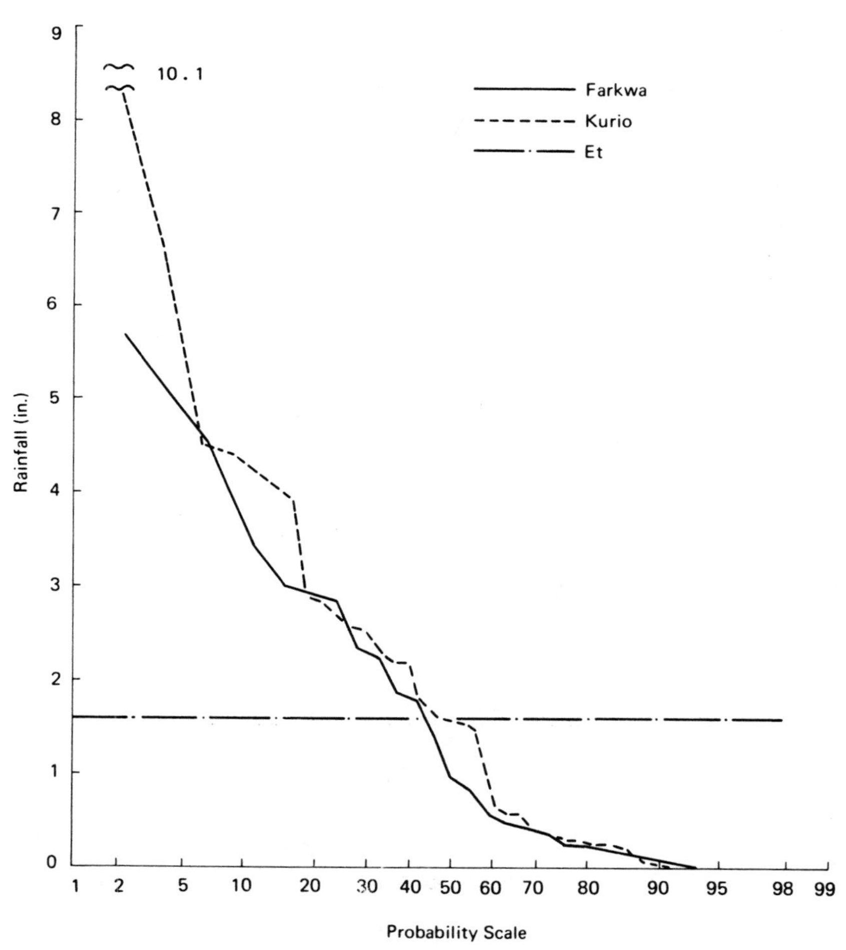

GRAPH 6 Rainfall probability for December III.

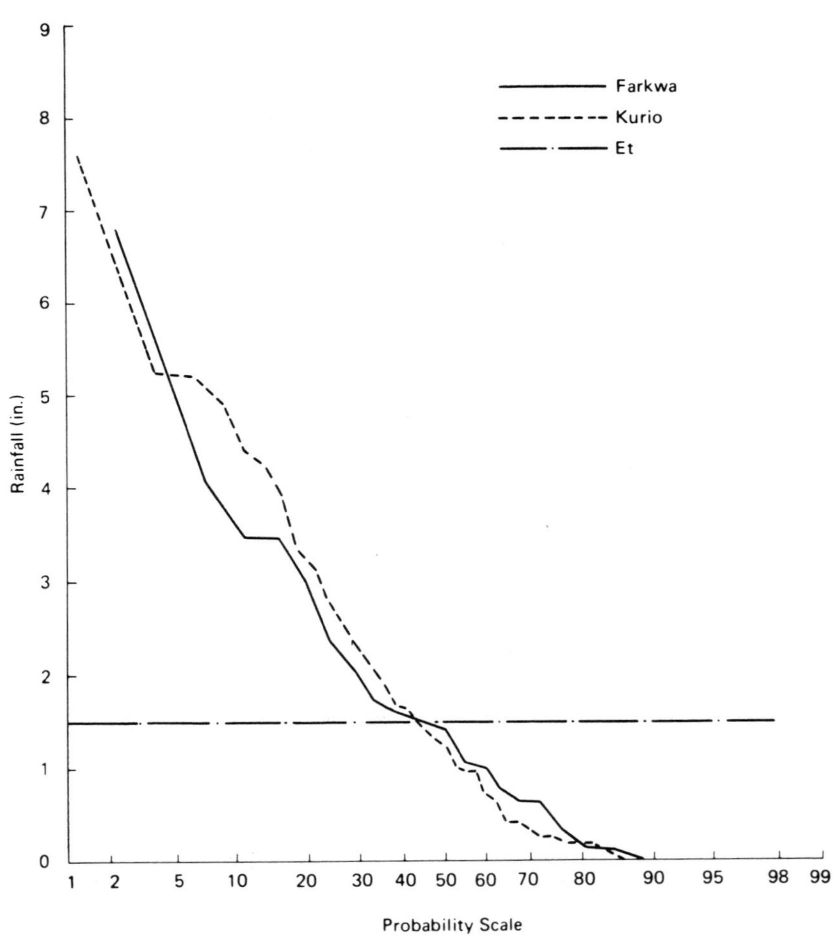

GRAPH 7 Rainfall probability for January I.

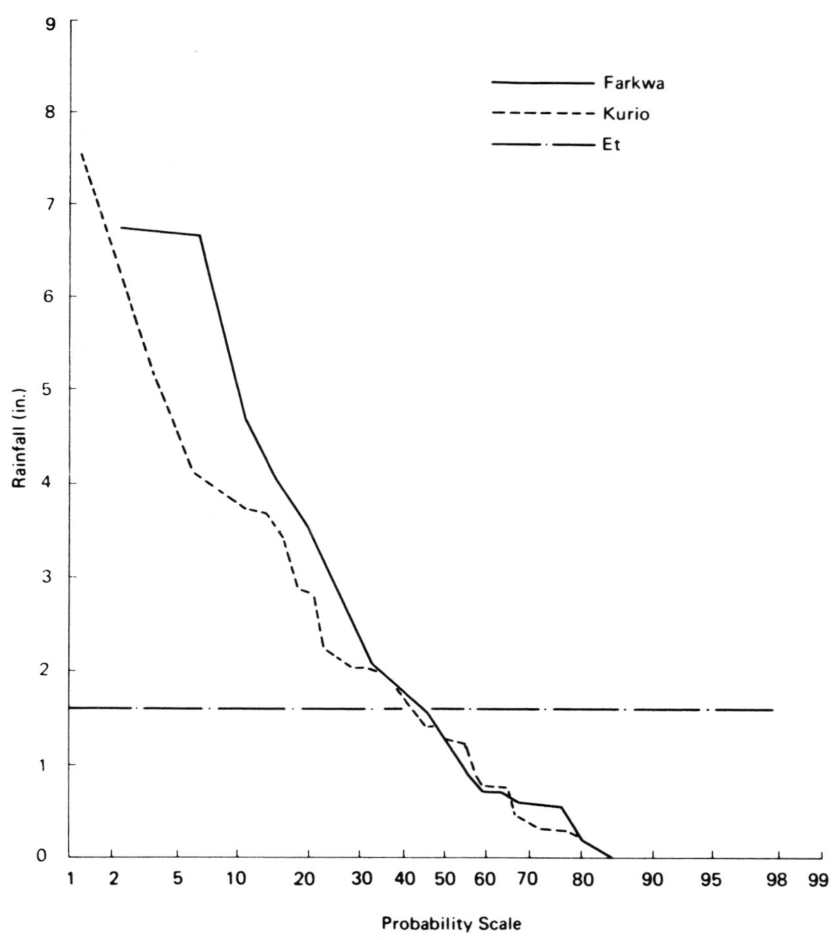

GRAPH 8 Rainfall probability for January II.

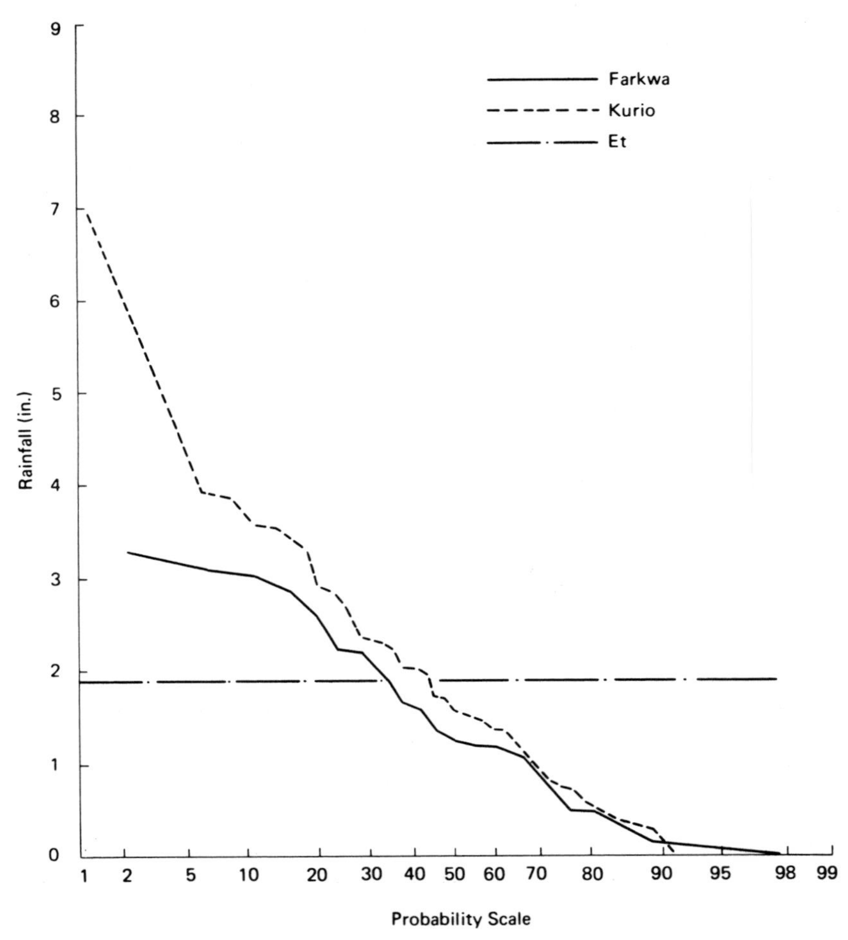

GRAPH 9 Rainfall probability for January III.

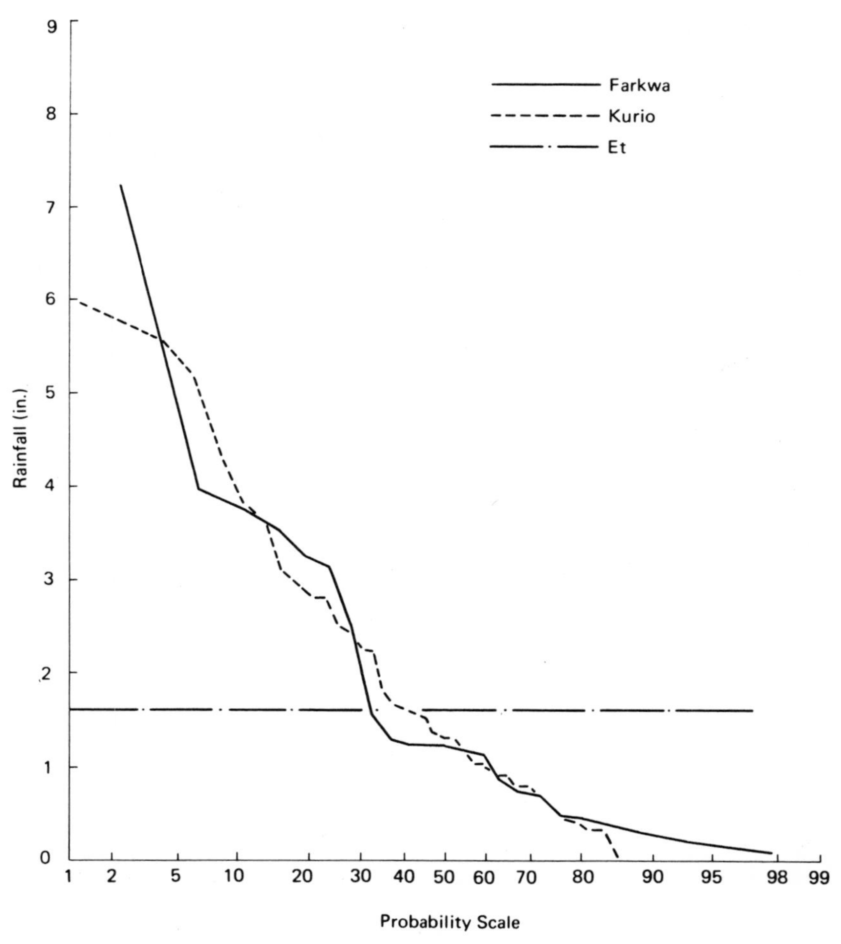

GRAPH 10 Rainfall probability for February I.

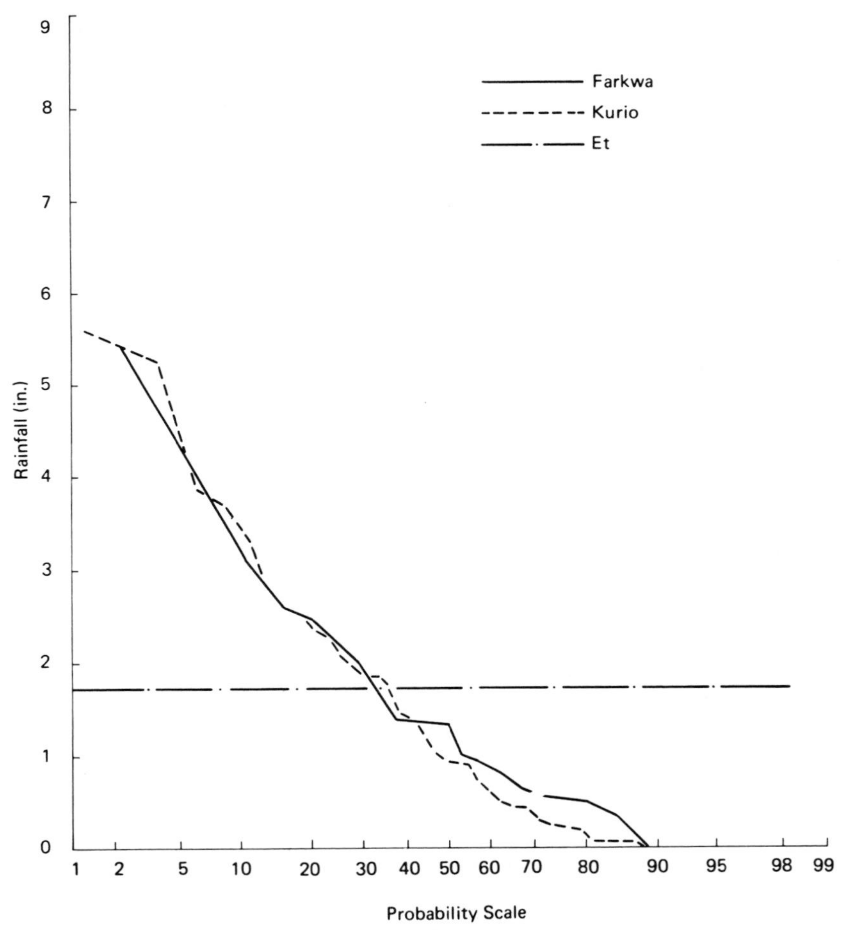

GRAPH 11 Rainfall probability for February II.

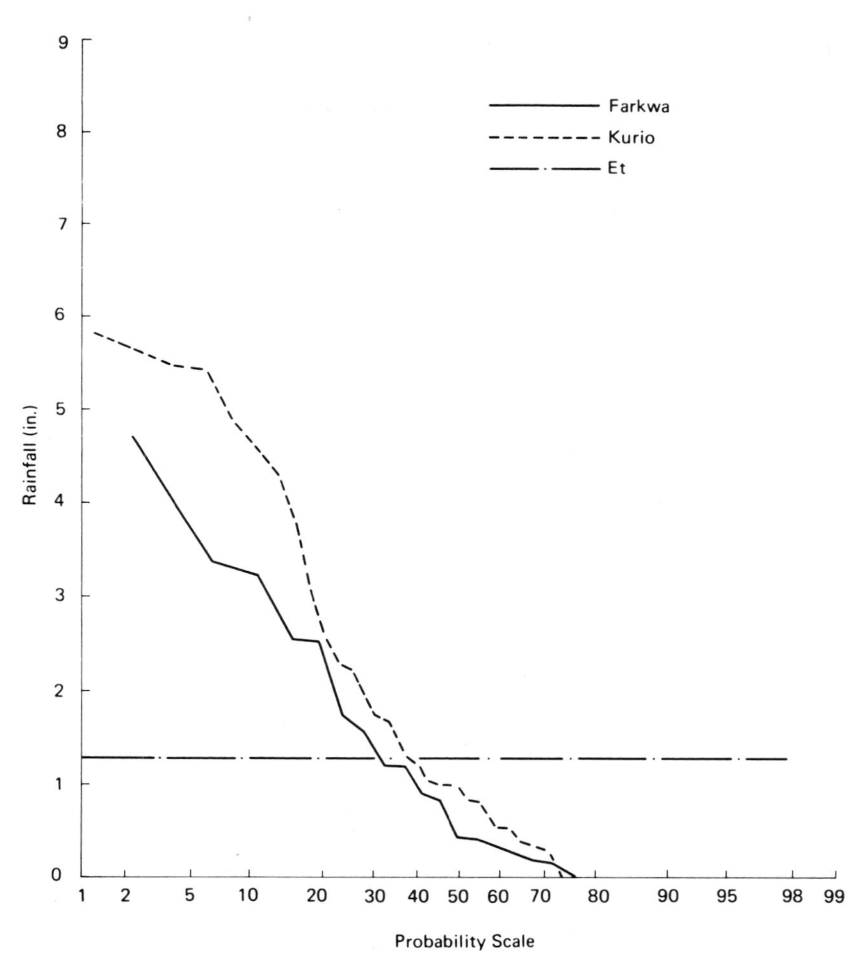

GRAPH 12 Rainfall probability for February III.

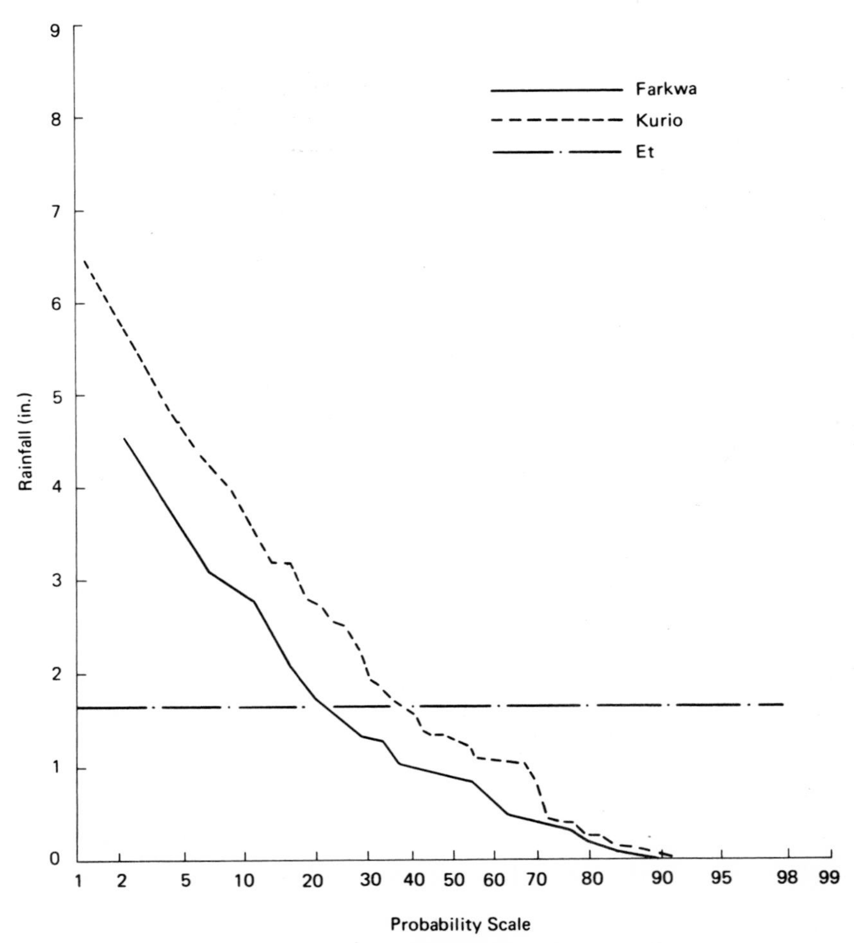

GRAPH 13 Rainfall probability for March I.

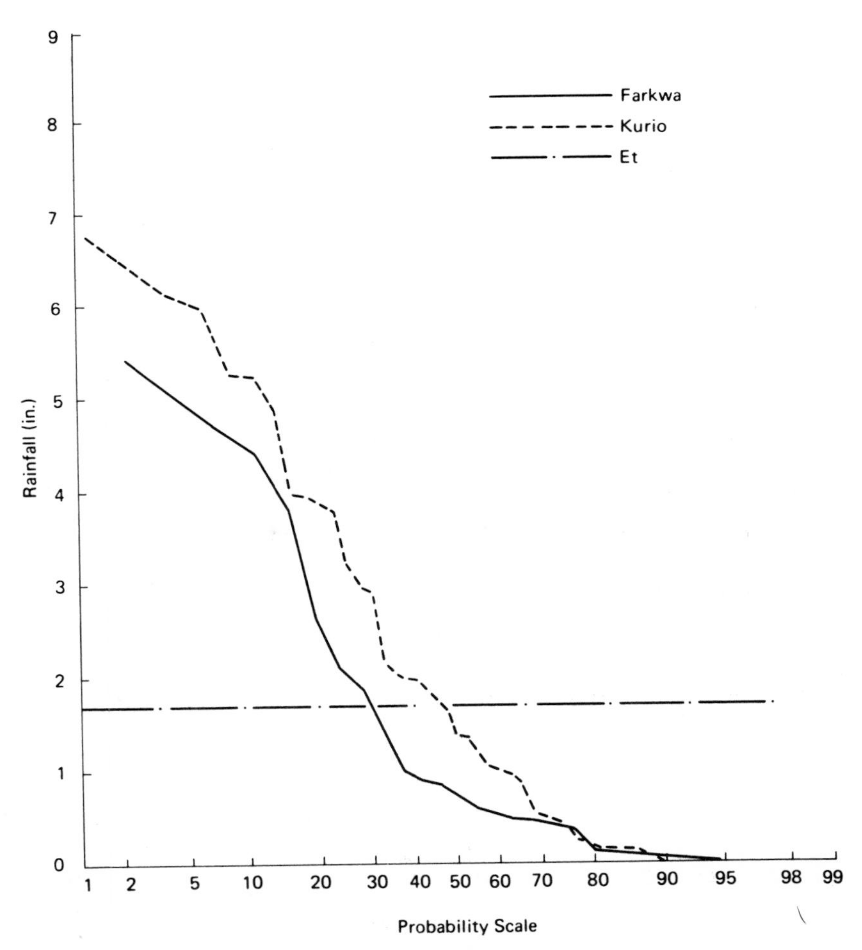

GRAPH 14 Rainfall probability for March II.

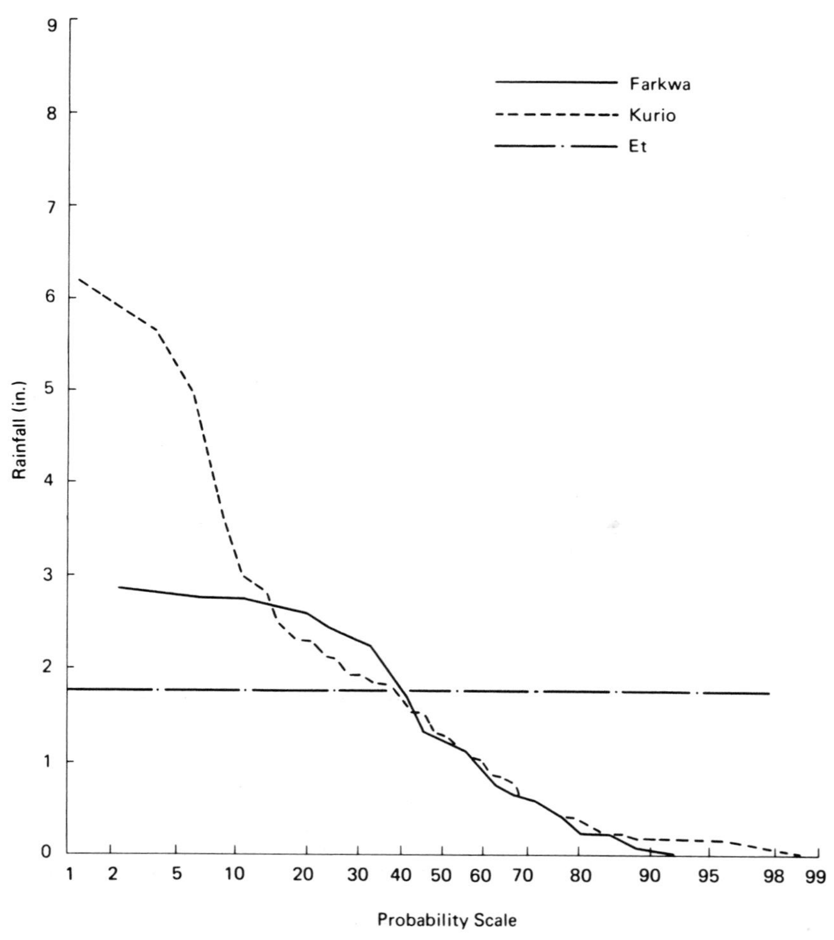

GRAPH 15 Rainfall probability for March III.

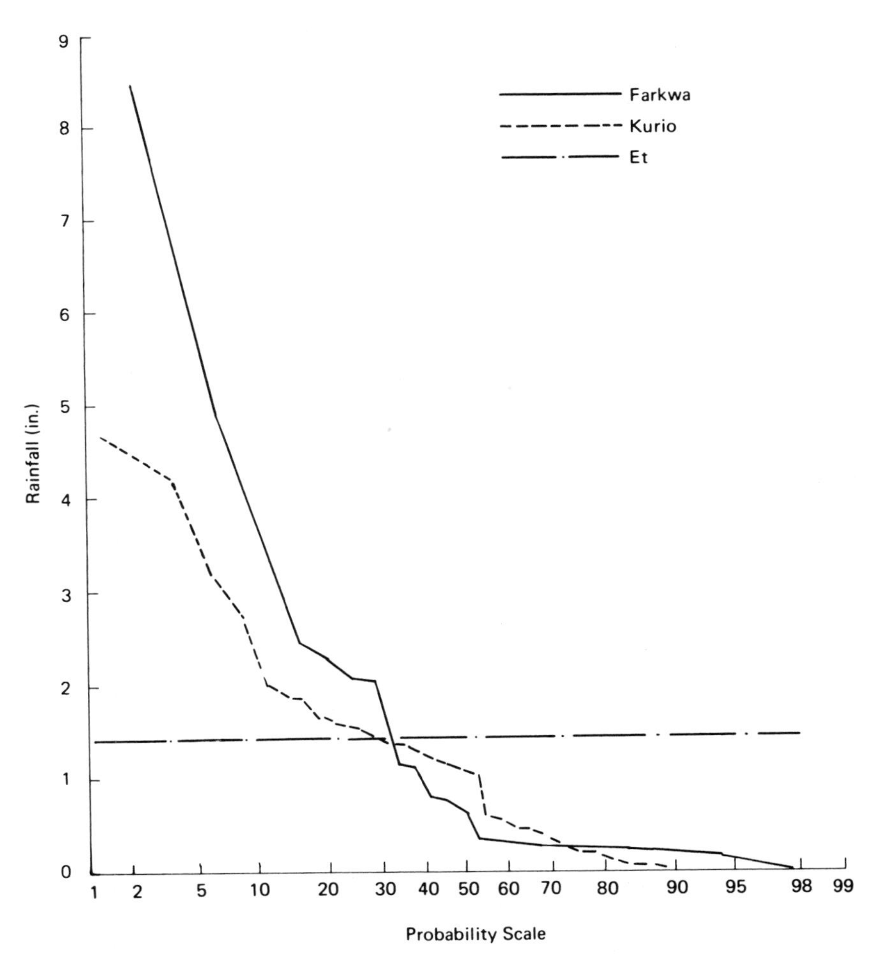

GRAPH 16 Rainfall probability for April I.

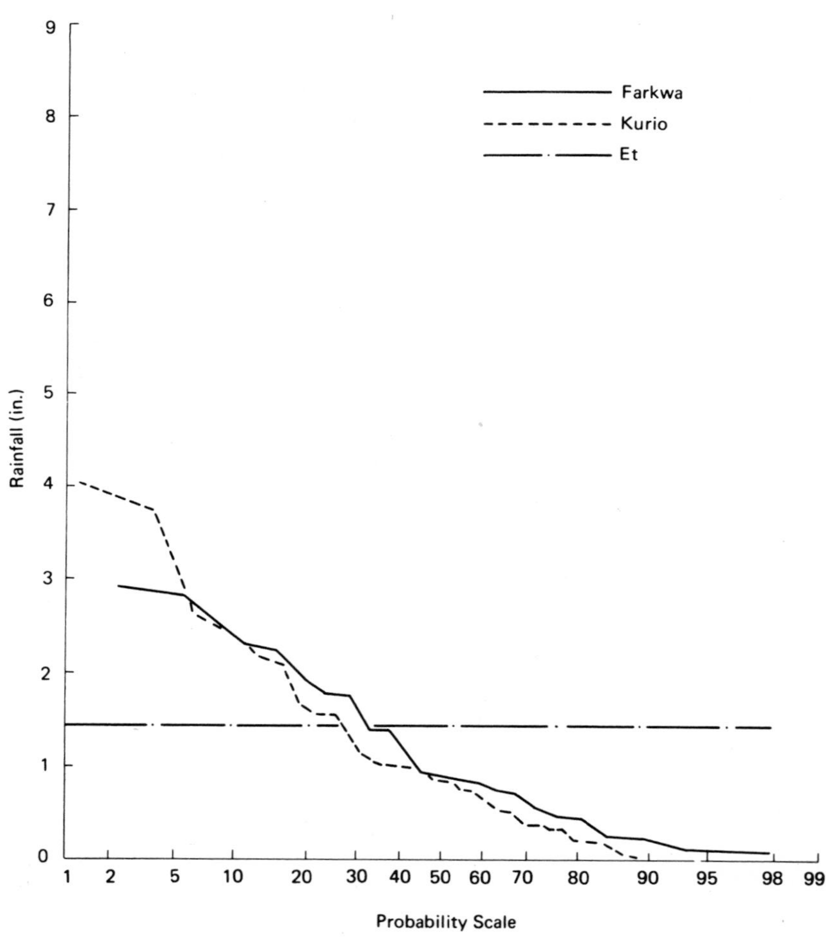

GRAPH 17 Rainfall probability for April II.

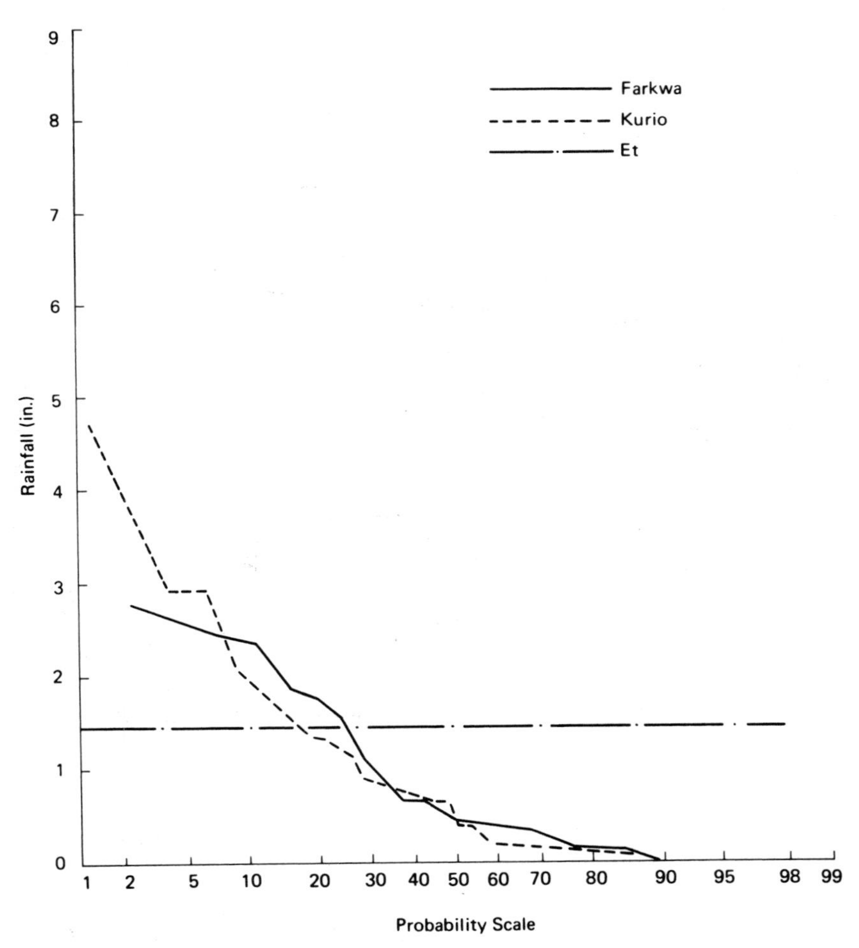

GRAPH 18 Rainfall probability for April III.

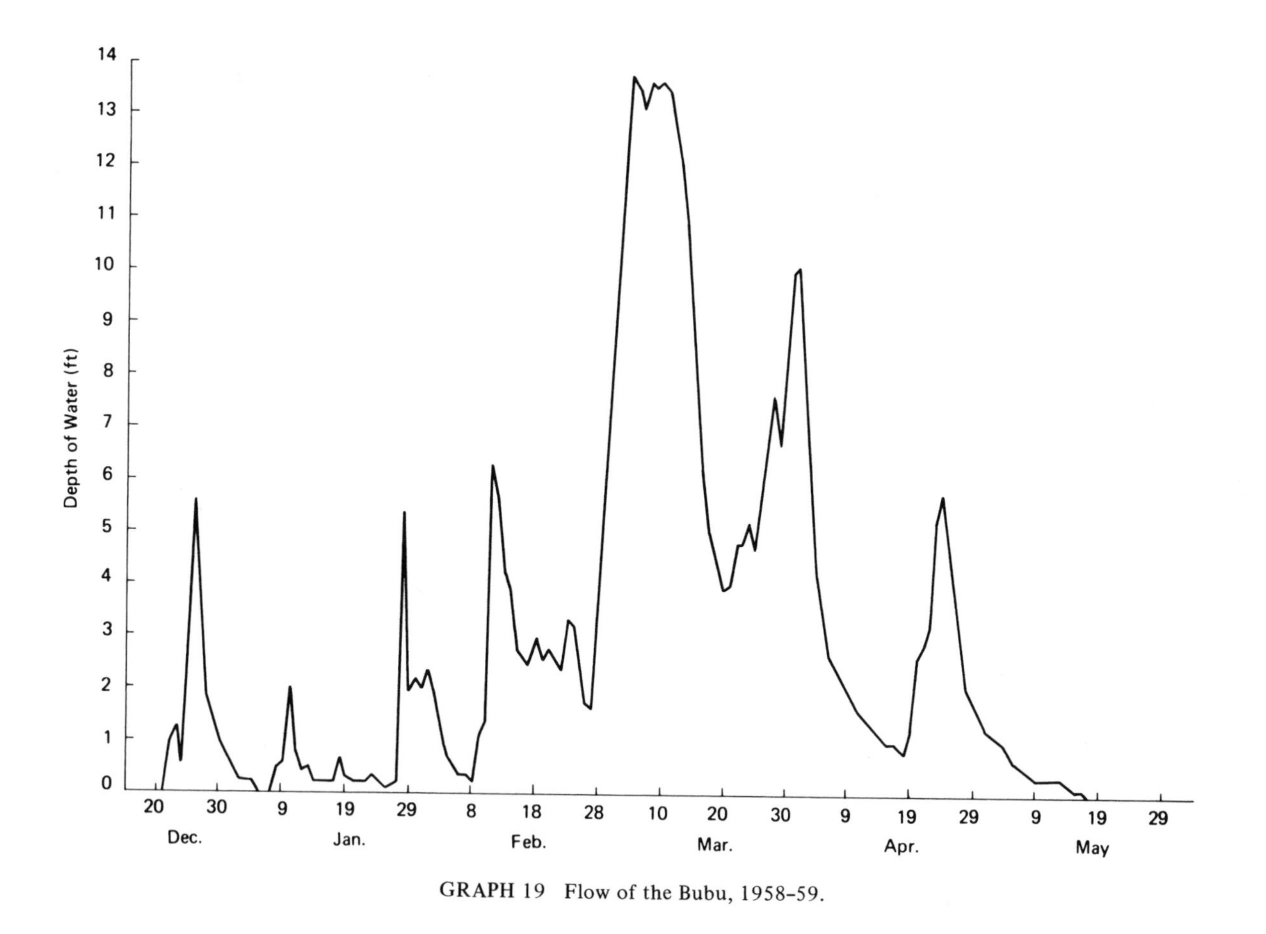

GRAPH 19 Flow of the Bubu, 1958–59.

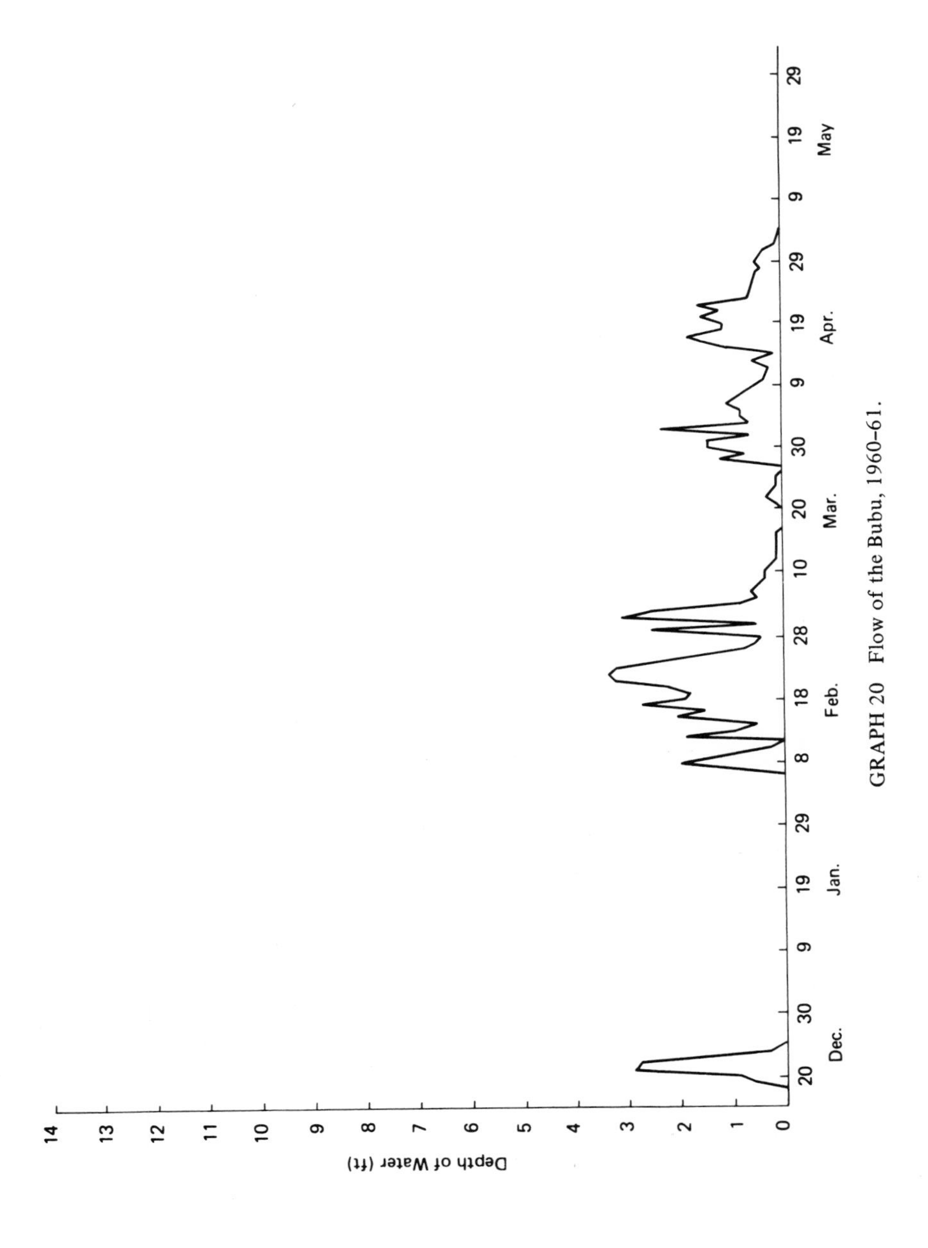

GRAPH 20 Flow of the Bubu, 1960–61.

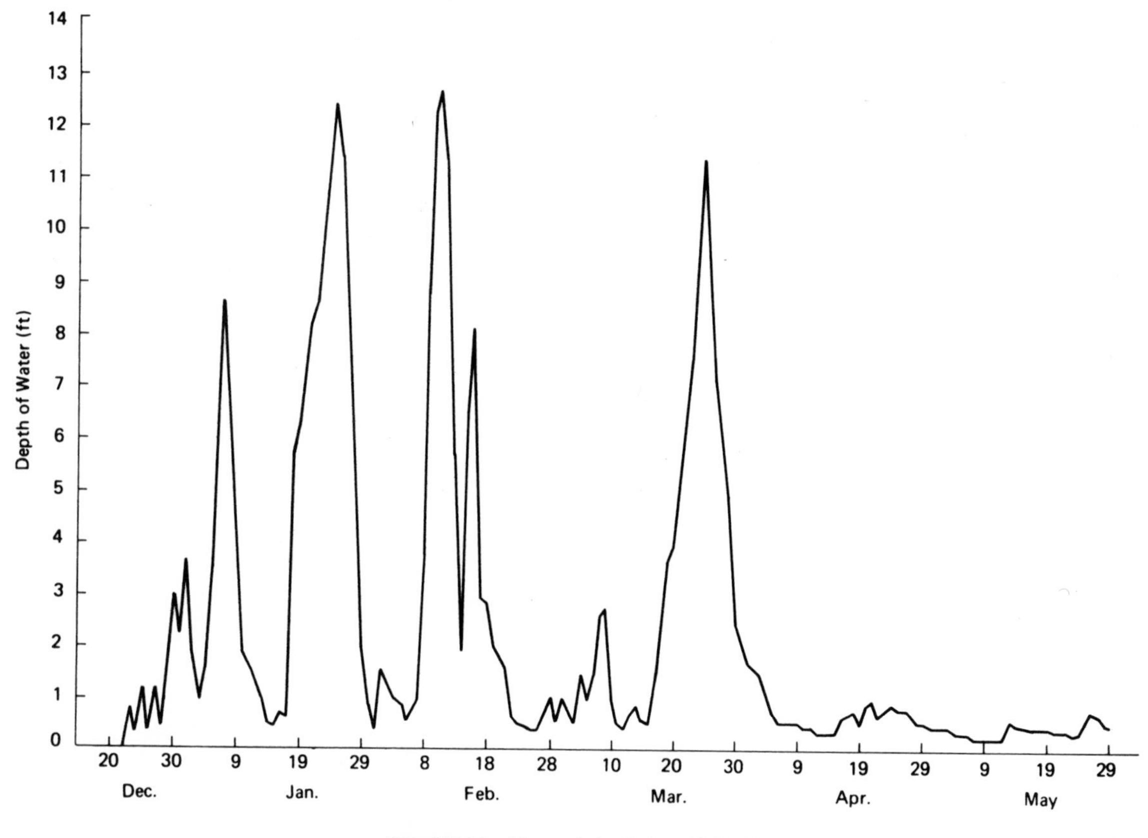

GRAPH 21 Flow of the Bubu, 1962–63.

CHAPTER 7

LIVESTOCK PRODUCTION

PRACTICES AND PROBLEMS

Most of the authorities who have commented on the economic development of Tanzania have stressed the importance of rationalizing the tending of livestock, particularly in the semiarid portions of the country. In fact, livestock is frequently cited as constituting one of the most (if not the most) potentially significant industries for the future economy.[1] The problems are considered to be essentially sociological ones, centering around the conservative attitude expressed by the majority of stockholders toward commercialization of their animals and the widespread overstocking that results. In comparison, we have seen that the Sandawe are not heavily committed to the "cult of livestock," and the countryside in general is not overstocked. Physical and economic problems appear to loom larger than those of culture.

The natural food resources for livestock in Usandawe are at best marginal. Heady has classified them in the *Hyparrhenia-Loudetia* simplex category, one falling outside his definition of the range land proper.[2] The basic limitation is the lack of grasses, both from a quantitative and qualitative standpoint. The dense tangle of the deciduous thicket country and much of the *Acacia-Commiphora* complex support only a very sparse undergrowth of grasses, and, in addition, hinder livestock movement. During the rains, a moderate grass cover springs up in the *Brachystegia-Combretum* woodlands, but by the end of June it is almost completely dried out. The grasses last longer in the openings of *Acacia-Commiphora* and on the *mbugas,* although even these are essentially finished by September. Actually, a good share of the food consumed by livestock derives from browse, notably the pods and leaves of *Acacia tortilis* and the leaves of *Terminalia sericea, Hippocratea buchananii, Barleria taitensis, Acacia senegal, Brachystegia spiciformis,* and *Combretum zeyheri.* The last two are used extensively in the few weeks preceding the rains when the woodland begins to produce flowers and leaves. A list of the main plants eaten by livestock and the season available is presented in Table 17.

From an examination of the seasonability of feeds, a roughly generalized grazing cycle can be constructed. During the rains and for a short time into the dry season the herds are concentrated in the *Brachystegia-Combretum* woodlands. After the crops have been harvested, grazing is allowed on the stubble (Figures 25, 26). People with few or no livestock normally allow others to graze their stubble in return for a future favor, such as milk, meat, or maybe even an animal. When this supply is finished, the herds live mainly off what they can find in the *Acacia-Commiphora* openings and on the mbugas. Then, just before the rains break, they are again taken into the woodlands. It must be re-emphasized that the Sandawe do not change their residence to adjust to the cycle of grazing. The animals are taken from and returned to the same kraal each day.

To the Sandawe, the real problem is water, not feed. Although the animals, particularly cattle, lose considerable weight during the dry season, they usually are able to find enough vegetable matter to keep themselves alive. In contrast, just about every stock owner interviewed admitted to losing at least several head of cattle each year because of lack of water. The very young and very old are especially susceptible, the critical period beginning about October and lasting until the heavy rains break. Few water holes large enough to meet the needs of livestock remain at the end of the dry season, and a round trip of 10 to 15 miles is not unusual. Heady has suggested that 3 miles is the maximum distance for optimum production.[3] Watering takes place every other day as compared to every day during the rest of the year. Boreholes have provided a welcome reliable water source for some herders, but there are too few to reach the bulk of the population. Another difficulty is that there is considerable overgrazing in the immediate vicinities of the available supplies. More often than not, though, this is due to a clustering of Masai rather than Sandawe.

Table 18 lists the diseases that could possibly infect livestock in Usan-

TABLE 17 Important Natural Fodder Plants of Usandawe

Botanical Identification	Sandawe Name	Usually Found	Season Available	Parts Used	Animals[a]
Grasses					
Aristida mutabilis	*segé*	*Brachystegia*	Rains–June	Leaves and stems	C
Cyndon dactylon	*zwaáré*	*Acacia-Commiphora* termite mounds	All year	Leaves and stems	C, S, G
Dactyloctenium giganteum	*helá*	*Brachystegia*	Rains–August	Leaves and stems	C
Sporobolus festivus	*girigitl'a*	*Acacia-Commiphora*	October–July	Leaves and stems	G, S
Panicum maximum	*belebéla*	Cut-over areas	Rains	Leaves and stems	C
Brachiaria serrifolia	*megembá*	*Brachystegia* and cut-over areas	Rains	Leaves and stems	C, G, S
Eleusine africana	*k'aran/ina*	*Acacia-Commiphora*	Rains	Leaves and stems	C
Chloris pycnothrix	?	*Acacia-Commiphora*	End of rains–July	Leaves and stems	C
Anthephora burttii	*kuntewa*	*Brachystegia*	All year	Leaves and stems	G, S
Harpachne schimperi	*xalá*	Termite mounds	January–April	Leaves only	C
Tragus berteronianus	*goró*	Widespread	April–June	Leaves and stems	C, G, S
Herbs and Climbers					
Disperma crenatum	*guláála*	*Acacia-Commiphora*	Rains–June	Leaves	C
Barleria taitensis	*búta*	*Acacia-Commiphora*	Rains–June	Leaves	C
Asripomea hyoscyamoides	*atha*	*Brachystegia* and cut-over areas	Rains–May	Leaves	C
Leucas stricta	*ts'iri*	*Brachystegia* and cut-over areas	Rains–April	Leaves	C
Solanum incanum	*intór*	Cut-over areas	All year	Fruit	G
Becium sp.	*goxomó*	*Brachystegia*	Rains	Leaves	C
Hibiscus lunarifolins	*kunguee*	Along streams	Rains	Leaves	C
Blepharis sp.	*gutl'gutl'*	*Acacia-Commiphora* and ant hills	Dry season	Leaves	G, S

TABLE 17 *Continued*

Botanical Identification	Sandawe Name	Usually Found	Season Available	Parts Used	Animals[a]
Shrubs					
Hippocratea buchananii	*hlenk'èta*	*Brachystegia*	Rains–September	Leaves	G, S
Ochna ovata	*kwelegi*	Steep hillslopes	October–November	Leaves and flowers	G, S
Premna sp.	*ts'andu*	*Brachystegia*	Rains	Leaves	G, S
Abrus schimperi	*mi/ísna*	Steep hillslopes and termite mounds	Rains–June	Leaves	G, S
Erythrococca atrovirens	*taráge*	*Brachystegia*	Rains	Leaves	G, S
Strychnos sp.	*tl'ikhaánka*	*Brachystegia*	Rains–July	Leaves	G, S
Vangueria tomentosa	*n!luúk'*	*Brachystegia*	Dry season	Leaves	G, S
Trees					
Brachystegia spiciformis	*inee*	*Brachystegia*	September–June	Leaves and flowers	C
Acacia tortilis	*áfa*	*Acacia-Commiphora*	All year	Leaves and pods	C, G, S
Maerua edulis	*segeléé*	*Brachystegia* and *Acacia-Commiphora*	November–June	Leaves	G, S
Dalbergia arbutifolia	*támba*	*Brachystegia* and steep hillslopes	All year	Leaves	G, S
Grewia platyclada	*xo'á*	*Brachystegia*	November–May	Leaves	C, G, S
Combretum zeyheri	*dérima*	*Brachystegia-Combretum*	September–June	Leaves	C
Dichrostachys cinerea	*dégera*	Cut-over areas and *Acacia-Commiphora*	All year	Leaves and pods	C, G, S

Afzelia quansenis	*dõ*	Widely scattered	September–November	Flowers	G, S
Diospyros fischeri	*manats'úk'a*	*Acacia-Commiphora*	All year	Leaves and pods	G, S
Boscia salicifolia	*zimbau*	*Brachystegia*	November–June	Leaves	C
Lonchocarpus eriocalyx	*kúmani*	*Brachystegia*	Rains–June	Leaves	C
Phyllanthus discoideus	*zíxaxa'*	*Brachystegia* and steep hillslopes	Rains–September	Leaves and pods	G, S
Premna angloensis	*parapapa*	*Brachystegia*	Rains	Leaves	G, S
Sclerocarya birrea	*án//uma*	Widely scattered	May–October	Fruit	G, S
Acacia mellifera	*rék'eto*	*Acacia-Commiphora*	Rains–June	Leaves	G, S
Acacia nilotica	*manange*	*Acacia-Commiphora*	All year	Leaves and pods	C, G, S
Markhamia obstusifolia	*hlohlobáá*	*Brachystegia* and cut-over areas	Rains	Leaves	C, G, S
Combretum langspicatum	*sóngalala*	*Combretum*	Just before rains	Leaves	C
Combretum collinum	*konkohlawa*	*Combretum*	Just before rains	Leaves	C
Terminalia sericea	*sen//a*	*Brachystegia* and cut-over areas	Just before rains	Leaves	G, S
Grewia holstii	*thenká*	Widely scattered	Rains	Leaves	G, S
Acacia senegal	*k'án!a*	Cut-over areas	Dry season	Leaves	G, S
Sterculia quinqueloba	*n!amahla*	*Brachystegia*	June, July	Leaves	G, S
Maerua tiphylla	*manuk'a*	*Brachystegia*	June, July	Leaves	G, S

[a] C = cattle; G = goats; S = sheep.

FIGURE 25 During the dry season, a common scene is livestock being driven to water (background). The donkeys are not part of the herd and are grazing on the stubble.

FIGURE 26 Sheep are not as numerous among the Sandawe as goats or cattle. They are valued primarily for sacrificial purposes and for oil obtained from the fat-tail.

dawe.[4] Of these, blackquarter is the most prevalent at the present time. On the monthly veterinary reports for Kondoa district over the past decade it has been mentioned more frequently as a cause of death than virtually all the others combined. The usual pattern is for several deaths to be recorded each month, with occasional more serious outbreaks. In February 1959, 42 deaths were reported among cattle throughout Usandawe and in January 1963, there were 32 cattle deaths from blackquarter in the Sanzawa area. As with all other diseases, the number of deaths that are unreported is a mystery. Anthrax is next in incidence of occurrence, followed by foot and mouth disease. An outbreak of the latter developed in

TABLE 18 Possible Livestock Diseases in Usandawe

Disease	Livestock Infected
Anthrax	C, G, S, D, Dg, P
Blackquarter	C, S
Brucellosis	C
Coccidiosis	C, Ch, Dg
Postular dermatitis	S
East coast fever	C
Ephemeral fever	C
Foot and mouth	C, G, S, P
Foot rot	S
Fowl pox	Ch
Haemorrhagic septicaema	C, G, S, D, Dg, P
Helminthiasis	C, S, G
Infectious opthalmia	C, S
Johne's disease	C, S
Listerellosis	C, S
Mange	C, S, G, Dg, D, P
Rift Valley fever	S, C
Rinderpest	C
Sheep pox	S, G
Streptothricosis	C, S, G
Tetanus	C, S, G, Dg, P
Trypanosomiasis	C, S, G, D, P, Dg

C = cattle; G = goats; S = sheep; D = donkeys; P = pigs; Dg = dogs; Ch = chickens.
Information courtesy of the Veterinary Division, Ministry of Agriculture, Dar es Salaam.

1965–66, causing a suspension of the livestock markets from September through January. Deaths seldom result from foot and mouth disease. East coast fever has been of minor significance in recent years, despite the absence of dipping stations in Usandawe. According to information gathered from informants, this portion of East Africa is apparently one that does not possess a high rate of natural infestation. A similar situation holds with respect to rinderpest, the great scourge of Masailand, and of much of Africa. An informed estimate places it as the number one livestock killer on the continent.[5] Since 1958, there has been no mention of rinderpest in Usandawe and the area is not included in the annual campaign of Kondoa district.

TSETSE AND TRYPANOSOMIASIS

Trypanosomiasis in livestock, the dreaded *nagana* of East Africa, is at the present time neither a regular nor a serious occurrence in Usandawe. The few scattered families who inhabit the tsetse-infested country keep no domestic animals other than chickens, while those who do happen to graze along its fringes—and these are primarily Masai, Gogo, and Turu—can obtain injections of ethidium bromide if needed. Essentially, the people and their livestock hold one area and the "fly" another. However, this pattern is a fairly recent phenomenon. First of all, the tsetse is a comparative newcomer to Usandawe. Second, the line of demarcation between the two basically incompatible life ways has not always been so sharp, and the history of contact between them has had important ramifications in the settlement and land-use geography of the area. It is our good fortune that since 1920 the movements of tsetse in central Tanzania have been fairly well documented by entomologists (notably C. H. N. Jackson and C. F. M. Swynnerton, the pioneers of tsetse ecology) and concerned administrators. Consequently, it is possible to discuss the interaction of man and fly with some precision.

The only major point to escape detection is exactly when the tsetse began their assault on Usandawe, but roughly the years around the close of World War I can be taken to give a good approximation. The Sandawe claim this as the time of entry and often in a jocular fashion will say that tsetse were "brought in by the English." An interesting indication from linguistics of the relative recentness of Sandawe acquaintance with tsetse is their use of the Swahili term *ndorobo* to denote them. There is no known Sandawe equivalent, whereas all other recognized insects apparently have an indigenous referent. The most conclusive evidence is that none of the prewar German sources mention the presence of tsetse in Usandawe. It is extremely difficult to imagine such an important element of the physical environment escaping their collective meticulous observations.

In any event, the flies were firmly established at Mangaloma in 1920 and were encountered at Serya during the same year (Map 8). Jackson located the origin for the advance in a colony on the upper Bubu at Berabera, which itself was derived evidently from the long established fly belt to the east below the rift escarpment on the edge of the Masai steppe.[6] By 1926 they had penetrated the Songa Hills as far as Gitl'au and were moving in a southwesterly direction across the old Kondoa-Singida road, now nothing but a track marking the boundary between

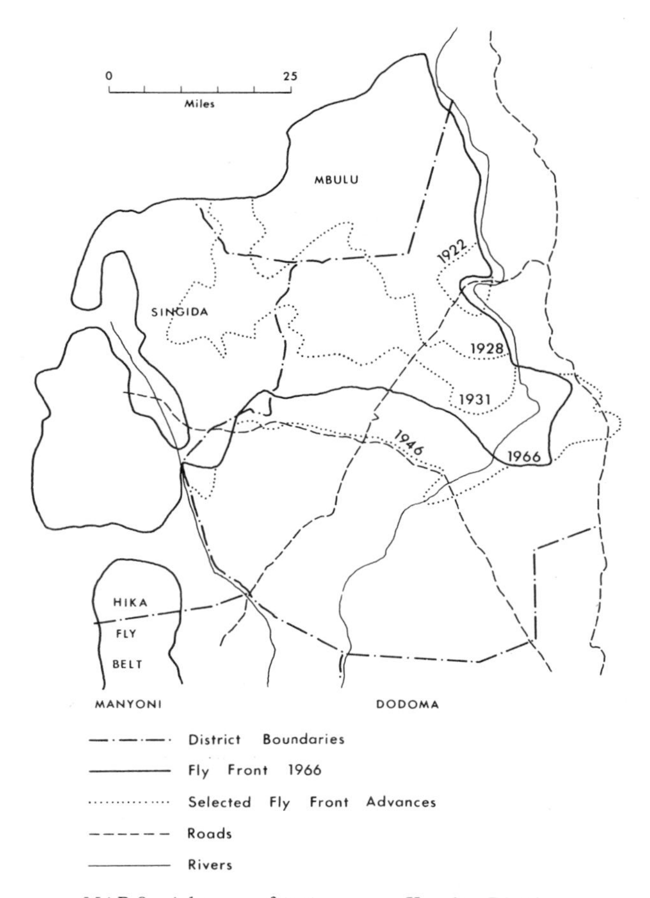

MAP 8 Advance of tsetse across Kondoa District.

the north of Usandawe and Mbulu district. The end of 1931 witnessed the spread of tsetse throughout the Songa Hills, while the front in the north and west reached about halfway to Kwa Mtoro and extended right across into Turu country up to the Mponde escarpment. Those families with livestock were essentially driven out in face of the onslaught.[7] The Barabaig retreated northward and henceforth were effectively shut off from Usandawe. More people began to leave the infested areas as cases of human sleeping sickness appeared.[8]

During the 1930's the advance continued. The road running west from Kwa Mtoro was reached and even crossed near Lalta, while a salient was pushed down the Mponde Valley to the south of Takwa. To the northwest, the flies had jumped the escarpment and were threatening to link up with the "great" western belt of *G. morsitans* located to the southwest of Singida town. However, by the end of the decade the authorities were hopeful that at least in Usandawe the invasion had reached its limits. The fairly dense human settlement south of the road between Kwa Mtoro and Ovada was counted on to hold the line here, and in the Mponde Valley the extensive stands of deciduous thicket were proving to be an effective natural barrier to penetration. Previous observations had shown that *G. morsitans* could not find suitable breeding grounds in such vegetation communities.[9] Similarly, the *Acacia-Commiphora* assemblage was supposed to offer the same kind of barrier, and consequently hold the tsetse in the Songa Hills at bay. But the flies proved more adaptive than the experts first thought, and about 1940 they again began to move westward toward the Sandawe Hills. By 1942–43 they had passed Ndoroboni and were at the outskirts of Kwa Mtoro. Farther south a bulge extended across the road to Farkwa and down the Bubu for a mile or so. It seemed as if the whole of the Sandawe Hills was about to be engulfed. With the exception of the Kwa Mtoro-Kurio-Ovada triangle, settlement was nowhere dense enough to prove an obstacle. Plenty of bush and stands of *Brachystegia* and *Combretum* woodland promised to make a comfortable home for *G. morsitans.* From here they could easily spread into the Chenene Hills and thence over much of Ugogo. Jackson, in fact, predicted that this would be the likely course of events.[10]

It was a frustrating time for scientist and administrator because there were no funds available to initiate positive action; they could only stand by and observe. The lone exception was a barrier clearing of some 13 square miles done by local inhabitants to the immediate east of Kwa Mtoro. Unfortunately, the area lacked readily available water resources and was not settled. The bush regenerated and the fly returned.

With the end of World War II significant remedial measures finally got under way. In November of 1945 a clearing scheme backed by government funds was implemented along the Bubu to alleviate the most serious threat. It took until December 1947 to complete and cost around $15,000, mainly for laborers' wages. The form of the clearing resembled a Y; the left arm and stem extending along the Bubu and the right arm along the Mkinke. Only the large trees, dominantly baobab and *Acacia rovumae,* were left standing. Also, a game fence about a mile long was erected across the Bubu

Valley near its junction with the Mkinke to discourage animals from taking this easy route into the Sandawe Hills. The main concern centered around elephants who, along with warthogs and kudu, serve as important hosts for *nagana.* The fence contains five strands of barbed wire fastened with pieces of tin which rattle in the wind. It is hedged with sisal on the east side. Finally, Sandawe settlers were brought in. An outbreak of *T. rhodesiense* occurred in the Songa Hills during 1946–48, with some 90 cases being reported each of the three years.[11] Most of the remaining families, about 350 in all, were rounded up and resettled on the clearings. Some Masai came in on their own because of the available water in the Bubu, Mkinke, and Zozo rivers. The borehole dug in 1955 at Poro brought in more Masai and as of 1966, tsetse had virtually disappeared from the lowlands, retreating to the Songa Hills.

More *Acacia-Commiphora* country was cleared between Kwa Mtoro and Iseke in 1948, followed by an extension up to the Mkinke for about a half mile on each side of the road in 1952. Once again, only baobabs and *Acacia rovumae* were left standing. Settlers were brought in mainly from the Kurio area where population pressure was mounting, though a few also came from Songa. Fly posts were set up, and are still in operation, at both Ndoroboni and Serya to de-fly vehicles passing through the hills. It is not very likely that cars, trucks, and buses would spread the fly belt, but a check reduces the probability of infected tsetse being transported into "clean" areas and biting livestock. A complete clearing of 5 sq mi was made at Takwa in 1957 (Figure 27). It was felt that the attraction of a permanent water supply would induce some families to move down from the intensively settled

FIGURE 27 Tsetse clearance near Takwa recently opened up a new area for settlement. The typical dispersed pattern has been established. Stretching away to the south is the peneplain of Ugogo.

Ovada area. To date the movement has been slow, with only about a dozen Sandawe households being established. Some Gogo have come in with their livestock and an Arab has several large maize fields, growing for the market in Singida. Earlier, in 1952, local effort cleared two and a half square miles around Handa. The final scheme envisioned for this particular part of Tanzania began in the vicinity of Mangaloma in 1965. Selective cutting was practiced to remove certain small trees and shrubs that continued research has shown are favored by *G. morsitans* for breeding.[12] This clearing is not meant for the Sandawe, but rather for Rangi around Kondoa town where overpopulation is critical.

Thus we are brought up to the present. Map 8 presents the line separating the region of human dominance from that dominated by the fly as of 1966. Though our knowledge of the reasons behind the movements of *G. morsitans* is still incomplete,[13] it appears that as long as the clearings are maintained by human settlement and not allowed to revert to bush, there is little chance of the fly again threatening Usandawe. In actuality, though, some Sandawe have begun to leave the eastern lowland clearings and return to the Songa Hills. The cluster of houses at Matonya began in 1964 and one or two families have moved in with each succeeding year. A few isolated homesteads have been set up as well. All settlement must take place north of the road, since the portion of the Songa Hills to the south was declared a forest reserve in 1947. The reasons behind the reverse movement are typically Sandawean. First, people have complained about cultivating the heavier soils exclusively and having to drink the fairly salty water generally found associated with the colluvial sediments. Furthermore, they know that honey and game will be easier to come by and, since most of the families have few if any large stock, it seems to make more sense to move back. The consensus, even among those who have so far remained behind, is that life is just much easier and more enjoyable "in the hills." Obviously, the situation requires careful watching as it pertains to maintenance of the clearings and to the possibility of human sleeping sickness reappearing.

CHAPTER 8

OTHER MEANS OF SUBSISTENCE

GAME AND HUNTING

Judging from historical trends, we can only conclude that hunting will continue to decline in importance among the Sandawe. The increasing reliance placed upon crop production and animal husbandry necessarily means that game will be farther and farther removed from the center of Sandawe life. In reality, it is not too difficult to foresee a situation arising similar to that found among the Turu, Gogo, Rangi, and just about all other East Africans, where hunting is confined to a very few individuals and where the vast bulk of the people receive little, if indeed any, contribution to their diets from wild meat. It was pointed out earlier how some of the younger generation are already well started on this path. Yet, looking at Usandawe in total perspective, there is no need for game to be considered as incompatible with crop and livestock development. They do not have to compete for the same land. As long as there are stands of woodland and thicket in the Sandawe Hills there should continue to be concentrations of dik-diks, duikers, bush pigs, klipspringer, hyraxes, and porcupines. These are all small and fairly prolific animals, requiring only limited expanses of "bush" in which to survive. Some sort of cropping of animals should be possible so that they continue to supply protein to the Sandawe. Perhaps a group of specialists or semispecialists might emerge to handle the hunting.

The real potential for game exploitation lies in the Songa Hills and on the adjacent plains and in the Mponde Valley, that is, basically within the now tsetse-infested region. Here can be found, in addition to the large numbers of small animals mentioned above, herds of eland, Thomson's and Grant's gazelles, and zebras, plus scattered bushbuck and greater koodoos. Add to this a superabundance of elephants and a situation exists that is ideal for the implementation of systematic hunting. The noted biologist Sir Julian Huxley has recommended that such places be set aside and used as "controlled hunting areas or managed for game-cropping" or else held in reserve for future use on the same lines.[1] Zambia has already begun to develop game reservations, apparently with some initial success.[2] As it presently stands, the tsetse-held region is essentially unproductive. To bring it into use for crops and livestock would require substantial investment in fly clearance, and then only very marginal land would be added to a pool already overstocked with marginality. In contrast, capital investment for controlled hunting would initially be much less, involving mainly expenditures on supervisory personnel, prevention of poaching, and perhaps the provision of supplemental water supplies.[3] In addition, there is an increasing body of evidence that indicates that wild animals, particularly the larger ungulates, are more efficient producers of protein in marginal habitats than domestic stock.[4] They generally can withstand drought better, are less susceptible to disease, and are able to convert the natural forage and grasses to a higher-quality product. Anyone who has eaten both types of animals throughout most of eastern and southern Africa will readily testify to the validity of the latter point. Then, too, it must be remembered that game animals have evolved an ecological balance. Different species live together exploiting different resources. Consequently, overstocking and overgrazing are less likely than with domestic stock. This is certainly an important consideration from the standpoint of resource management.

Game animals have another value in addition to protein, that of enhancing tourism. The Songa area could never be expected to attain the stature of places like Ngorongoro, Manyara, and the Serengeti, for it lacks the enormous concentrations of animals found in these famous parks and possesses a rugged terrain to hinder easy accessibility. But it might very well attract those who think the existing parks are too tame and commercialized, in other words, the tourists who want to see and discover things more or less on their own.

Another attraction of the Songa Hills is the numerous rock paintings scattered about, many of them not yet officially documented. The scenes are normally found on exposures of smooth granite and depict in the main various animal and human forms, though much of an apparently symbolic nature also is encountered.[5] These works of art constitute one of the most fascinating antiquities of East Africa and deserve more attention and publicity than they have received. At this time, some of the sites are in danger of being permanently damaged by the fires of illicit timber cutters and honey collectors. Granite overhangs provide cozy temporary shelters.

In conclusion, one can make a strong case for preserving populations of game animals in Usandawe. As F. Fraser Darling, one of the world's leading authorities on wildlife conservation, has put it, to exchange many wild animals, particularly the ungulates, for livestock and, I might add, cultivation, in areas which are distinctly marginal for such uses is to destroy a valuable natural resource and a "marvelous ordering of nature."[6]

BUSH PRODUCTS

The maintenance of areas for wild game means the preservation and perhaps establishment of essentially undisturbed stands of vegetation. But game preservation is only one benefit of "natural" plant assemblages. Probably their most significant function centers on watershed management and erosion control. Certainly, there should be no repetition of the "lunar landscape" produced in Irangi. There, virtually all the vegetation has been stripped from the hilltops and steep slopes, with the result that severe gullying has cut into the light-textured upland soils, while correspondingly the drainage channels have become choked with the sediment washed down from above. To the detriment of Usandawe, one can now observe a similar process in its beginning stages at several locations, notably around the mission stations. At both Kurio and Ovada, despite manuring and permanent cultivation, population densities seem to have locally exceeded their maximum under the existing economic framework of the country, while at Farkwa people have attempted to cluster and still employ the old extensive methods of land usage. The vegetation is being rapidly denuded. We shall return to the matter of population carrying capacity in Chapter 9.

Additionally, woodland and thicket provide an array of products for local consumption. A previous section indicated the extent to which the Sandawe have traditionally exploited such resources, and it seems that as long as they are available many bush products will continue to be utilized. In house construction there is no indication that burnt bricks and metal roofs are about to sweep the countryside. Most people seem quite well content with the comforts and conveniences of the *tembe*. It is a fairly waterproof and windproof dwelling, requiring little expenditure other than the labor to build it. New sections are easily tacked on should the size of the family expand.

There are many different timber species employed in *tembe* construction, depending basically on how much effort an individual wants to put forth and how long he wishes the house to last. The softer, more easily worked timbers endure for about 4 to 5 years, while the heavier and harder varieties can last up to 10 years or so. Of the latter, the more common include *Acacia nigrescens, Berchemia discolor, Boscia mossambicensis, Bussea massaiensis, Cassia abreviata, Dalbergia melanoxylon, Dichrostachys cinerea, Grewia mollis,* and *Vitex payos.* Other woods widely used in the construction of household utensils, ax and hoe handles, furniture, musical instruments, etc., include *Ostyroderus stuhlmanni, Albizzia tanganyicensis, Afzelia quanzensis, Dalbergia arbutifolia, Commiphora caerula, Commiphora ugogensis, Entandrophagma bussei, Markhamia acuminata, Premna angloensis, Pterocarpus angleoensis, Sclerocarya birrea,* and *Strophanthus eminni.* The barks of *Adansonia digitata* (baobab) and *Brachystegia spiciformis* provide fibers for the making of moderately durable ropes. For greater details on the constructional uses of plants, see Appendix C.

It is not likely that commercial exploitation of wild vegetation will develop in Usandawe. A few valuable tree species are present, but they are widely scattered and are not found in appreciable quantities. The most valuable is undoubtedly *Dalbergia melanoxylon*, the so-called East African "black wood." It is prized in Europe for the construction of small, high valued articles like canes, musical instruments, and inlays.[7] *Afzelia quanzensis* has been much sought after for furniture, doors, and formerly dhow making, while *Premna angloensis* and *Phyllanthus discoideus* are reputed to be fairly good woods for furniture and cabinets.[8] Mention also should be made of *Acacia senegal*, a source of gum arabic. The Sandawe formerly tapped these trees for sale to the local Arabs, but no longer do so because of its fall in value. However, the major problem facing woodland development is not the species composition, for this can be altered, rather it is the sparse rainfall. Growth rates for trees are severely retarded, making widespread exploitation unprofitable. Though no precise figures are available, it is fairly apparent that the quantity of biomass production would be insufficient even under the most scientific management.

Available foods from uncultivated plants also must be considered in a discussion of production potential. Though they cannot match the overall yields of domesticated crops, many of them have nutritional qualities that merit consideration. The importance of several of the small herbs in providing the main source of green vegetables for the diet has already been commented upon. Similarly, a good share of the fruit intake is derived from bush sources, and in an area where the moisture regime is unfavorable for the successful establishment of most planted varieties and where, because of transportation and purchasing-power problems, importation on a large scale is not feasible, it becomes imperative that at least selected traditional species remain at hand. Studies dealing with the comparative nutritional value of bush fruits are vitally needed. Perhaps some might prove to be domesticable.

In addition, it is necessary to inquire into seasonal rhythms in the availability of the various bush foods. This becomes an especially pertinent consideration for the period just prior to the harvest, when the crops reaped the previous year will be at their lowest ebb and when, therefore, any dietary supplements will be welcome. As Table 19 shows, though peak productivity comes somewhat earlier, there are still some foods to be had during the critical time. In fact, every month can provide at least two or three ready species. It must be remembered that we are speaking about most years, for as we have seen, it is highly unlikely that bush foods are of much value during occasions of severe drought.

Whether any of the medicinal plants are of special value remains to be determined. At the present time, the Sandawe still rely on their own remedies more than on those available in local dispensaries and hospitals. Only the most widely known and used remedies are included in Appendix C. There are undoubtedly many more, especially those that are coveted by various medicine men. Individuals who are interested in the subject of African medicinal sources are referred to the impressive volume entitled *Medicinal and Poisonous Plants of Southern and Eastern Africa*, by J. M. Watt and M. G. Bryer-Brandwijk.[9]

The production of honey and beeswax could be considerably expanded. Swarms of honeybees, *Apis mellifera adansoni*, exist and, though sometimes difficult to manage because of their viciousness, they are prolific producers. In the deciduous woodland, shortly before and during the rains, they thrive on pollen from the flowers of *Brachystegia, Lannea,* and *Combretum* species,[10] and the *Acacia-Commiphora* country also provides some good flowering plants for nectar, especially during the dry season. Evidently the deciduous thicket communities are not as attractive. There is virtually no flowering in the dry season and even during the rains, at least from my casual observations, the frequency of swarms seems to be appreciably less than in the other two vegetation communities, a general comparison confirmed by local informants. The Itigi thicket has the added problem of lacking surface moisture from which the bees can drink. Water is made

TABLE 19 Seasons of Availability for the Main Bush Foods in Usandawe

Plant	Use	Season Available
Adansonia digitata	Pulp and seeds eaten plain	All year, depending on age of tree
Berchemia discolor	Fruits eaten plain	March through May
Boscia mossambicensis	Fruits eaten plain	November and December
Brachystegia spiciformis	Seeds of pods dried, boiled, and then eaten plain	January through April
Bussea massaiensis	Seeds of pods ground and mixed with water before eating	August through November
Canthium burtii	Fruits eaten plain	April and May
Ceratotheca sesamoides	Leaves boiled and eaten as a relish; a Swahili *mlenda*	All year, because of storage
Cissus trothae	Fruits eaten plain	February and March
Corchorus trilocularis	Leaves boiled and eaten as a relish; a Swahili *mlenda*	January through April
Cordia ovalis	Fruits eaten plain	February and March
Cordia rothii	Fruits eaten plain	Feburary and March
Cyphostemma knittelli	Fruits eaten plain	January
Delonix elata	Seeds boiled and eaten	June through October
Erythrococca atrovirens	Seeds boiled and eaten	February through April
Ficus fischeri	Fruits eaten plain	April and May
Ficus hochstetteri	Fruits eaten plain	April and May
Ficus sysmorus	Fruits eaten plain	June through August
Grewia bicolor	Fruits eaten plain	April and May
Grewia holstii	Fruits eaten plain	April through November
Grewia mollis	Fruits eaten plain	March

Grewia platyclada	Fruits eaten plain	June and July
Grewia similis	Fruits eaten plain	March and April
Gyandropsis gynandra	Leaves boiled and eaten as a relish; a Swahili *mchicha*	All year, because of storage
Haplocoelum foliolosum	Fruits eaten plain	March and April
Hydnora johannis	Pulp eaten plain	December through May
Kedrostis hirtella	Fruits eaten plain	January through April
Lannea floccosa	Fruits eaten plain	March through May
Lannea stuhlmannii	Fruits eaten plain	March through May
Maerua edulis	Fruits eaten plain	January through April
Momordica rostrata	Fruits boiled, then eaten	All year
Neorantenenia pseudopachyriza	Bean pods boiled and eaten	March
Opilia campestris	Fruits eaten plain	February and March
Peponium vogelii	Fruits eaten plain	January through October
Pouzolzia parasitica	Young leaves boiled, then eaten	December and January
Sclerocarya birrea	Nuts eaten plain	April and May
Sesamum angustifolium	Leaves boiled and eaten as a relish; a Swahili *mlenda*	All year, because of storage
Strychnos innocua	Fruits eaten plain	November through January
Tamarindus indica	Seeds of pods ground and mixed with water before eating	August through November
Tapiphyllum floribundum	Fruits eaten plain	April and May
Vangueria acutiloba	Fruits eaten plain and dried	All year, because of storage
Vangueria tomentosa	Fruits eaten plain and dried	All year, because of storage
Ximenia americana	Fruits eaten plain	All year
Ximenia caffra	Fruits eaten plain	November through February
Ziziphus mucronata	Fruits eaten plain	November and December

available in the settled areas throughout the year from the wells, water holes, springs, and boreholes used by humans and livestock.

In contrast to most tropical things, various diseases and pests do not appear to be noticeably prevalent. The depredations of the honey badger constitute probably the most serious threat. Some problems are caused by moths and beetles that breed in the honeycombs and also by the predations of ants and wasps. Apparently all of these can be fairly easily controlled.[11]

A ready-made local market exists in the surrounding towns of Kondoa, Singida, and Dodoma, where the lack of nearby thick woodland does not permit the existence of many honeybees. Africans, by and large, crave honey as a sweetener and for use in *pombe*, and there is little question that demand is far in excess of supply. At present, the Sandawe market mainly beeswax from their hives. Just about all the honey is consumed locally, very little finding its way to the towns. The present volume of production cannot satisfy both demands.

The main impediments to increased production lie in the realms of technology and finance. The traditional cylindrical hives cannot be managed properly. They contain numerous leaks and cracks and have no queen excluders. Whether a swarm will actually colonize a specific hive or not seems basically a matter of chance, since nothing is done to encourage the bees, like baiting the hive or providing nearby water supplies. Furthermore, the crude methods of gathering the honey and wax destroy many of the young bees and larvae, which discourages the swarm from returning. During the late 1940's a scheme was set up to remedy a few of these problems and thereby promote the development of honey and beeswax in Kondoa district. Unfortunately, it was realized more on paper than in practice. According to Sandawe informants, all they were shown was how to construct bark hives, a procedure that did not make much sense to them because such hives require more labor than the traditional log ones and last only through one rainy season. This is uneconomic! "Why not build two or three log hives, get as much, if not more, total honey and wax and have them last for a couple of years?" they ask. What obviously is needed is for the complete technology and potentiality of modern beekeeping to be demonstrated and explained by trained supervisory personnel. Initially, financial assistance, especially for the hives and houses, will have to be provided to interested individuals, since it is unlikely that anyone will have the necessary starting capital.[12] The Sandawe are eager and skilled honey hunters, and it would be a shame to waste a human as well as a natural resource.

FISH

Fish are taken from four sources in Usandawe: the Bubu, Mkinke, Mponde, and Zozo rivers. Only the former is of any real significance, and though it is possible to catch fish just about anywhere along its banks, the main fishing area is below the rapids near Kologa. Here the two dominant species in the Bubu—*Clarias mossembicus* (barbel) and a *Labeo* species, probably *L. cylindricus*—congregate as they are washed down by flood waters from above or as they swim up to the barrier from below.[13] The main fishing season is restricted to April and May when the water is back in the channel of the river but still a foot or so deep. During these two months the Sandawe come from miles around to fish. Many of those who live near the stream can be seen every day for two or three weeks. Some outsiders, particularly Rangi, also come. When the floods are in full spate little or no fishing is done, since Sandawe techniques are not able to cope with very rapid flows of water. After June there are only a few scattered pools of water containing small numbers of emaciated specimens. Occasionally, very young children can be found fishing at such spots.

It will be useful to summarize the various methods employed by the Sandawe in fishing. Following Hickling, we can divide these into individual and multiple methods; that is, whether one or more than one fish is taken at a time.[14]

Of the individual methods, the most elementary is manual collection. When the flow of the water in the river is reaching its ebb, the women wade in with conical-shaped wicker baskets (*kibwabwa*) open at both ends. The procedure is simply to plunge the basket over a sighted fish and then to remove it through the top by hand. Simple hand-catching is also employed. Young children attempt to grab fish stranded in pools with their bare hands, while women utilize a piece of cloth wrapping. A fish spear (*mósóma*) is used by men on larger barbel. It is normally aimed at a fish rather than thrust randomly. In contrast is the use of the gaff hook (*mokhobé*). This is the most popular method of fishing by the men during the height of the season. They wade about in the water swinging the hook from side to side in hopes of snaring medium-sized barbel. The hook, line, and pole method has been introduced quite recently, but is not widely employed, being found mainly among young boys. Earthworms are put on the hook as bait.

Multiple methods include traps, barriers, nets, and seines. By far the most widely used is a nonreturn trap (*mogono*) constructed in the shape of a funnel. The fish enter the wide end and once inside cannot find their way out again. The trap is usually placed between rocks or held in place by branches at some spot where there is only a narrow path through which the fish can pass. Branch barriers (*lutángo*) are constructed across the stream either to ensnare fish or else to force them to cluster behind the barrier, from which they can be removed easily by hand. A modern innovation is the netting of fish. The net is strung across the channel and supported by two poles. The nets are not made locally, but purchased in Dodoma. Their comparative recentness is attested to by the fact that only the Swahili term *wavu* is used in reference to them.[15] The final Sandawe fishing method is that of the basket seine (*mogono*). Its construction is similar to the nonreturn trap, though without the funneled entrance at one end. It too has a handle on top that women hold onto as they move it through the water where clusters of fish are known to be. No poisons seem to be used, nor do the Sandawe fish at night with lights.

Some fish are brought to market in Kondoa and Dodoma, but by far the largest share are consumed locally, where they constitute another valuable source of protein for the Sandawe diet. Like many bush fruits, fish are doubly valuable in that the main bulk comes at the time shortly before harvest when it is most needed. Drying and smoking allow for a part of the catch to be preserved for up to 3 months. Whether increased commercial exploitation is possible depends on the quantity of fish available and the spread of more efficient fishing methods such as netting. So far no assessment has been made on how intensively the Bubu can be fished; at the present level of activity, there do not seem to be any indications of exhaustion. As with honey, market demand is well in excess of supply.

CHAPTER 9

POPULATION-LAND BALANCE

It would seem appropriate to conclude the discussion of subsistence potential with an assessment of how many people Usandawe can support. Is the area overpopulated and, if not, how close is it to the margin? The question has already been broached in the analysis of the Sandawe's hunting and gathering past and has been alluded to on several other occasions. The time has now come to try to make somewhat more precise calculations.

Immediately we are confronted with a serious problem. No one has yet devised a way of measuring absolute human carrying capacity.[1] All the available empirical estimates are technologically relative and thus give population estimates applicable only to the particular technology under investigation. What is more, these measures are rather crude and unsophisticated, being based on very broad and general categories of data and requiring a considerable amount of guesswork. Nevertheless, they can be useful in giving us at least a rough idea of the population–land balance under existing conditions in Usandawe.

For present purposes, the best approach appears to be the one formulated by Allan.[2] It springs from long experience by its author in Zambia under environmental and cultural circumstances very similar to those encountered in Usandawe. The more widely known methods of Conklin and Carneiro were found to be inapplicable.[3] The former is predicated on a system of shifting cultivation, while the latter presupposes a pattern of village settlement. As we have seen, not all the Sandawe are shifting cultivators and settlement is dominantly of the dispersed homestead type.

Three initial measurements are required before proceeding with Allan's computations: the cultivation factor (C), the cultivated area needed to support one person; the percent of land in the area that is fit for cultivation (P); and the land-use factor (LU), defined as the total number of "garden areas" or fields required to keep one field continuously under cultivation.[4] For example, suppose a field can be cultivated for 3 years before abandonment is necessitated and that 12 years of resting is needed to restore soil fertility. A farmer would then have four fields in various stages of fallow plus one under cultivation, giving a land-use factor of 4 + 1, or 5. The above three variables are combined in the formula $100 \text{ LU} \times \frac{C}{P}$ to determine the minimum acreage for carrying one person, which allows easy conversion into the population per square mile.

Computing the acreage per person is readily done by sampling, and for Usandawe on unmanured lands it works out to be about 1.2 when all fields are included. The cultivation factor presents a more perplexing problem, one that Allan has called the "most puzzling and intractable" of all.[5] Allowances must be made for rock outcrops, permanent swamps, seasonal waterlogging, local steep slopes, ironstone and hardpan formations at or near the surface, especially sterile soils, and the maintenance of bush for purposes such as sacred groves and rain shrines.[6] There is no standard method of combining these phenomena. All that can be done is to make educated estimates from maps, air photos, and, above all, familiarity with the place. For Usandawe, I have estimated 25 percent for the hills and 35 percent for the *xáts'a* lowlands, figures that are not markedly different from Allan's own in Zambia.[7] The land-use factor also becomes a problem without long-term observation of the agricultural cycle, but according to informants 3 to 4 years of cultivation followed by 15 to 20 years of fallow on the upland soils seems to be about average. It should be remembered that the Sandawe do not practice systematic fallowing, so that one has a right to be skeptical of their opinions on this matter, though the sample soil profile from *//'eú* regrowth (Table 9) tends to support the estimation. There is no general consensus about the *xáts'a* soils because of the Sandawe's lack of experience in cultivating them. A rough estimate based on a few samples indicates that 5 years of use followed again by 15 to 20 years of fallow is probably not greatly in error. The respective land-use factors are then 6 and 4.5.

Combining the above values in the equation, we find that under shifting cultivation the light upland soils are capable of supporting 21 persons/sq mi and the *xáts'a* soils 41, with an overall balance for Usandawe of around 30. These figures can be compared with the actual present-day density of 15/sq mi. To get a more realistic view of the population–land balance, the focus must be reduced to the Sandawe Hills and their immediate environs, in other words, to the area where the vast majority of the people are located. Here the density jumps to 35/sq mi. But now a new consideration has to be introduced, namely the more intensive land-use practices employed in the vicinities of Kwa Mtoro, Kurio, and Ovada (Figure 28). To account for it, all the variables in the equation need to be modified.

The biggest change comes in the land-use factor, which drops to around 2. Incomplete planting of the grain fields because of insufficient manure and the continuance of shifting cultivation on the groundnut, Bambarra nut, sweet potato, and bean fields keeps it from attaining a value of 1. With increased fertility from manuring, the total acreage requirement per person is lowered to near 1. Finally, the land is somewhat flatter than throughout most of the Sandawe Hills and in consequence the cultivation factor can be raised to 30 percent. The resultant maximum desirable population density works out to 95/sq mi. The actual density is around 100, and thus we can speak of a problem of overpopulation in this area, a fact seemingly borne out by the increasingly serious gully erosion. A summary of the various factor values is provided in Table 20.

Across the remainder of the Sandawe Hills the density is only about 17/sq mi, though local concentrations as at Farkwa and Sanzawa do rise above the optimum limit for the shifting cultivation practices in vogue.

FIGURE 28 The most densely populated portion of Usandawe is located between Kwa Mtoro and Kurio, producing a cultivation-steppe landscape. Gully erosion can be seen encroaching from the left.

TABLE 20 Population–Land Balance Values in Usandawe

	Cultivation Factor (C)	Percent Cultivable Land (P)	Land-Use Factor (LU)	Maximum Density
Upland soils (unmanured)	1.2	25	6.0	21
Xáts'a soils (unmanured)	1.2	35	4.5	41
Manured soils	1.0	30	2.0	95

The precision of the above figures, of course, can be contested, given the assumptions of the equation, but they help to confirm quantitatively the general impression that, except for localized circumstances, Usandawe is not beset by critical problems of overpopulation. If all the land now held by tsetse—some 531 sq mi—were cleared and the population redistributed fairly uniformly across the countryside, even utilizing only the light upland soils under shifting cultivation, more people could be supported in Usandawe than are at present. With a shift to permanent, manured tillage, over four times as many could be carried.

There are several reasons why the population has not expanded up to its limits, two of which stand out as particularly important. First, there is the traditional desire of the Sandawe to live in dispersed settlements and to maintain as much intervening bush as possible (Figure 29). Second, there

FIGURE 29 Dispersed patrilineal–patrilocal homesteads are found over the greater share of Usandawe. The father's *tembe* is demarcated by the adjoining cattle enclosure.

are evidently preferable alternatives to remaining in Usandawe. Out-migration appears to indicate that the people have made their own assessment of the land's carrying capacity. Relative to opportunities elsewhere, Usandawe is already overpopulated. The empirical equations present only theoretically attainable values that are isolated from a larger context and only peripherally related to individual desires and wishes. Migration data more closely touch on these facets and, as Hunter has shown, there should be more work on bringing them into the assessment of the population–land balance problem.[8]

IV

CONCLUSION

CHAPTER 10

CONCLUSION

The topic of change and development is an immense one, requiring for its understanding contributions from as many viewpoints as possible. By reference to a specific case example, this study has attempted to explore what insights an ecological orientation might be able to lend. Hopefully, the presentation has been stimulating enough so that criticism will lead to the formulation of new and incisive questions and to the production of more precise research theory and methods. To this end, it would seem useful to present a brief summary of the main findings.

Establishing the rate and direction of past change was the first step. Because of the lack of written records before the last decade of the nineteenth century, much of the information had to be of an inferential nature, some of it even negative in character, yet the correspondence appeared sufficient enough to show with a fairly high degree of probability an increasing reliance by the Sandawe on stock-keeping and crop cultivation at the expense of hunting and gathering. It was also demonstrated how initially this change seems to have been gradual and was stimulated by widening culture contacts and subsequent population growth. The pattern fits Fried's category of adaptation, as opposed to either annihilation or incorporation.[1] Had the movement of Turu, Barabaig, or Gogo been swift, violent, and in great numbers, the Sandawe undoubtedly would have ceased to exist as a distinct people either by being physically destroyed or by being incorporated into a larger society. As it happened, Sandawe culture was able to survive, albeit in modified form, and even to absorb some of the migrants.

More recently, change has continued to operate under the influence of contacts with differing cultures and expanding population. However, instead of being purveyed by informal, essentially nondirected neighbors, change is now carried on by a more purposeful government and commercial economy. Taxes, hunting regulations, and the like have necessitated modifications in Sandawe subsistence practices. So far, adaptation still seems to be the main process, though incorporation is inevitable in the long-run drive to establish a Tanzanian national unity.

The second step centered on measuring the available natural resources—precipitation, ground and surface water, soils, vegetation, fauna—and their present pattern of exploitation. The main purposes were to contribute to the formulation of an ecological approach to changing subsistence systems and to establish a base for assessing the potential for future resource development, thus helping to meet urgent planning needs. It is to this second consideration that I would like now to turn, directing my remarks to the general course of change I see as most logically dictated by the data.

What strikes one most about Usandawe is, in Huxley's words, the brittleness of the environmental system.[2] A delicate balance seems to have been struck among the various physical phenomena, one that allows very little leeway for human error. Severe damage can quickly result from overstocking and overcultivation, an observation amply testified to in surrounding areas and increasingly in Usandawe itself. The obvious conclusion is that hasty, badly planned actions will probably not succeed.

A concomitant of environmental brittleness is marginality for human habitation. Crop production in particular rests on shaky foundations. An unreliable and limited moisture supply, infertile soils, pests, and diseases—all constitute obstacles to production. With sufficient effort, pests and diseases probably can be controlled fairly effectively. Similarly, fertilization could be carried out to upgrade the soils; but even then, added complications appear. In areas where there is a likelihood of drought occurring during the growing season, the strong, early growth of plants stimulated by fertilization places a heavy burden on crop-water supply.[3] No matter how we view it, there seems to be no solution to the moisture problem.

Looking beyond the provision of a basic food supply, cash cropping encounters further difficulties. Even if garden crops, such as tomatoes and onions, could be grown successfully, there is no central urban population concentrations such as those available to the Rangi in Kondoa, the Turu in Singida, and the Gogo in Dodoma to absorb the production. Additionally, the lack of all-weather roads in Usandawe makes access to these outside markets exceedingly difficult. On the world export scene, only groundnuts and fire-cured tobacco seem to be at all feasible, but it is doubtful if even these could be grown on a large scale. Neither is in particularly strong demand.

Considering all the impediments to establishing a sound crop-based economy and the numerous investment avenues needed to overcome these difficulties, it would seem advisable to keep cultivation within its present areal limits. To let it expand would only add to the magnitude of the problem. Indeed, contraction is preferable if the necessary out-migration can be absorbed elsewhere. However, a redistribution of population on any large scale is not likely in the foreseeable future. Consequently, plans should concentrate on improving productivity in the better areas, with feasible modifications including upgrading manuring practices, developing more rational crop and field rotations, expanding tie-ridging, utilizing surplus storage facilities, and perhaps introducing ox plowing. Under existing conditions, it is probably best not to plan beyond providing a local food supply. The increasing dependence on maize at the expense of more drought-resistant grains should be halted.

Livestock ranching faces fewer obstacles and thus holds out more potential for commercial exploitation, particularly on the *Acacia-Commiphora* lowlands and the mbugas. The Tanzania Agricultural Corporation ranch at Kongwa in Ugogo has shown that ranching schemes can succeed under carefully regulated conditions.[4] Initially, several important additions must be made in Usandawe. Many more boreholes need to be sunk, more facilities for disease control should be provided, and probably fencing should be supplied. Bush clearance with a view toward improving feed availability is also an important consideration. A fair amount of research has been devoted to finding suitable planted pastures. So far the most promising grasses for semiarid environments include *Cynodon plectostachyum* (star grass), *Chloris gayana* (Rhodes grass), *Cenchrus ciliaris* (African foxtail), *Pennisetum purpureum* (elephant grass), *Pancium maximum* (Guinea grass), *Panicum makarikiensis, Urochloa pullulans,* and *Dicanthium papillosum.* As for legumes, *Stizolobium deeringianum* (velvet bean) and *Glycine javanica* seem to be most promising.[5] Market demand is on the rise and promises to continue to do so, especially within a wider East African context. Furthermore, the Sandawe constitute an important human resource. As we have seen, they have not yet become overly conservative about livestock and they certainly demonstrate a capacity for absorbing change. If a scheme is set up to demonstrate the advantages of sound livestock ranching, I see no reason why the Sandawe should not respond. They have learned from example many times in the past.

Cattle in any number are best kept out of the Sandawe Hills. Pasture establishment and regulation would be more difficult than on the lowlands and there is the danger of erosion along tracks of steep slopes. Sheep and goats are feasible, however. They are able to convert rough browse fairly effectively into an acceptable product so long as their population does not get out of hand. The possibilities of introducing more scientific methods of chicken-raising should be explored.

I have already indicated my opinion that tsetse-infested country should be left essentially as it is, exploiting it mainly for hunting, bush produce (especially honey), and tourism. If the surrounding *Acacia-Commiphora* country is cleared for ranching, there should be no concern about the fly spreading beyond its present limits. The same policy is recommended with regard to the Itigi thicket. There is no immediate need or apparent advantage in clearing it for human settlement, and in its present state it acts as a barrier to the western *Glossina morsitans* belt. In addition, the thicket offers a good laboratory for plant geographers and ecologists. Both areas can be thought of as holding land in storage for future occupance if population pressure becomes critical.

The plea, then, is for an integrated, multiple land-use pattern that is adjusted to the resource base and to cultural and market considerations and that minimizes human suffering and dislocation. As far as I can detect, there are no compelling arguments for a radical, overnight transformation of the Sandawe and their land. Rather, time and care must be taken in order to avoid what René Dumont would call another "False Start in Africa."[6]

APPENDIXES

APPENDIX **A**

SELECTED ITEMS OF SANDAWE MATERIAL CULTURE

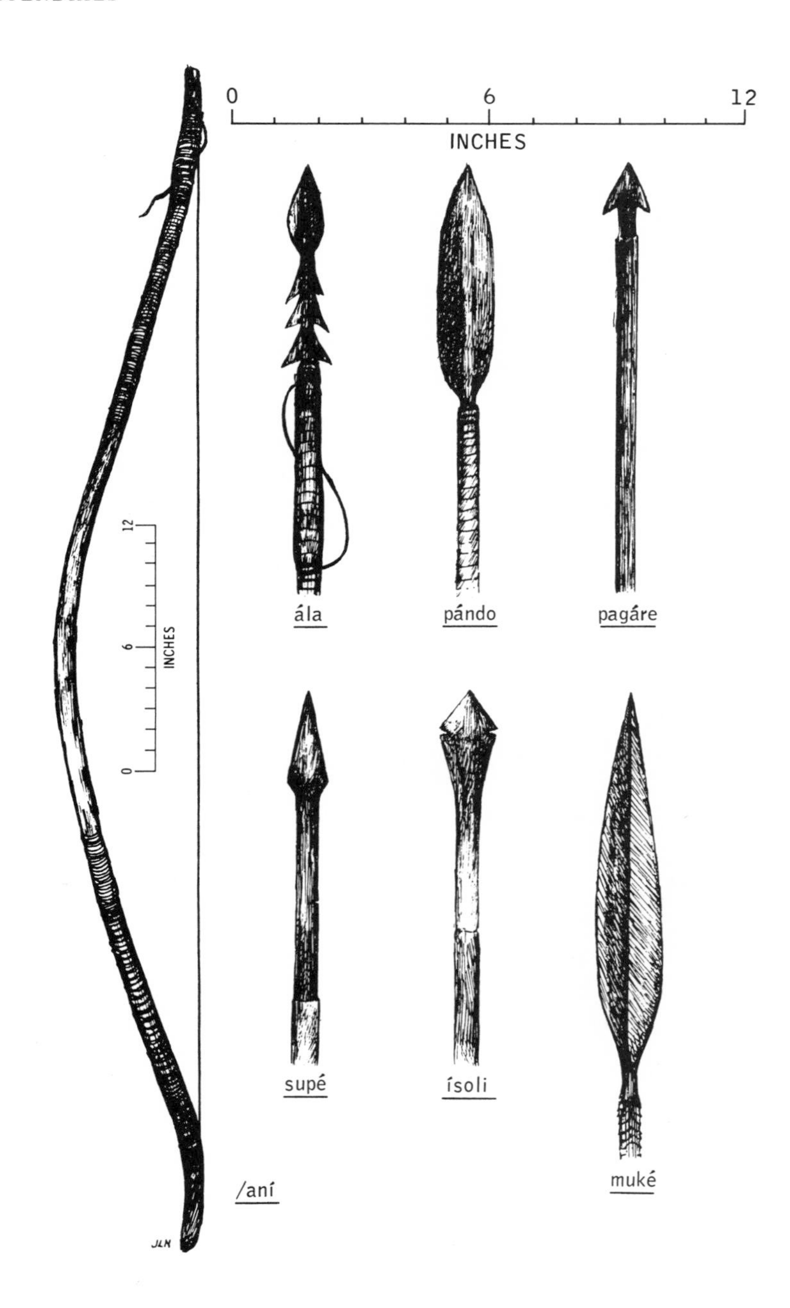
0
6
12
INCHES
0
6
12
INCHES
ála
pándo
pagáre
supé
ísoli
muké
/aní
JLN

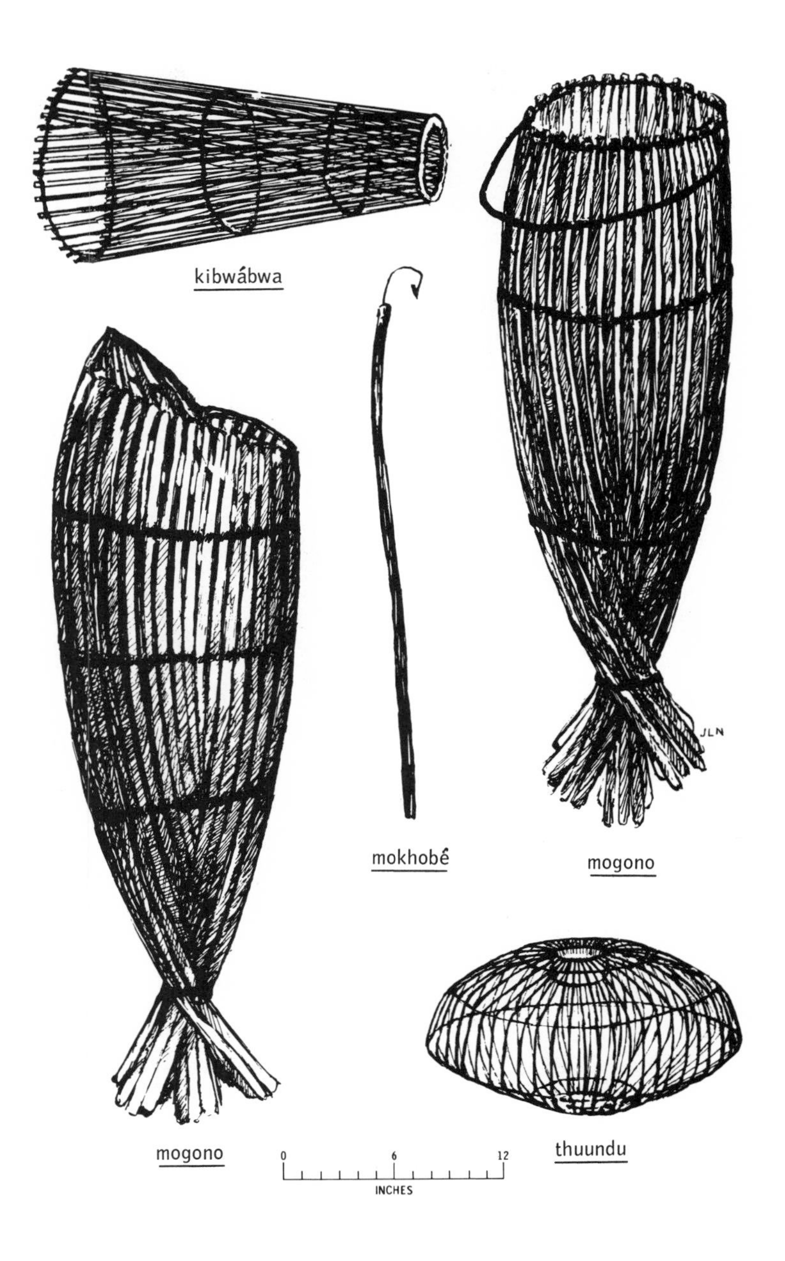
kibwábwa
mokhobê
mogono
JLN
mogono
thuundu
0
6
12
INCHES

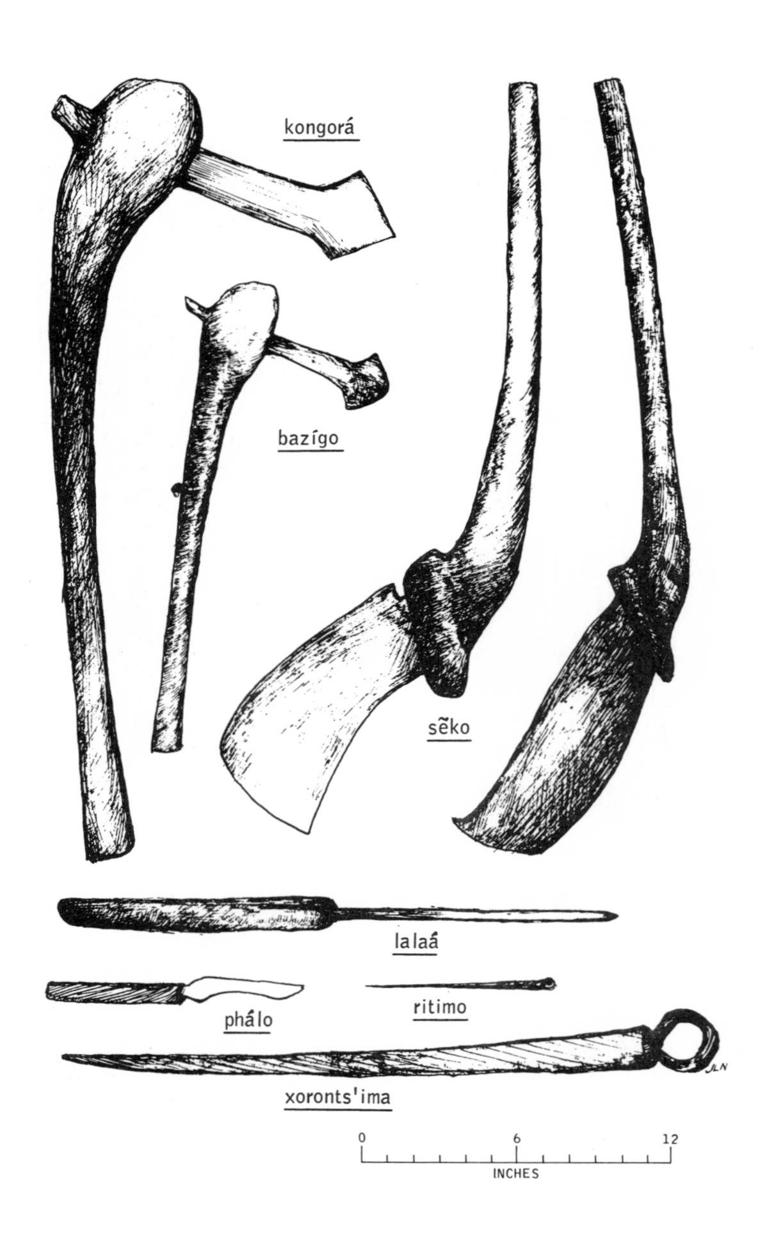
kongorá
bazígo
sẽko
lalaá̂
phálo
ritimo
xoronts'ima
JLN
0
6
12
INCHES

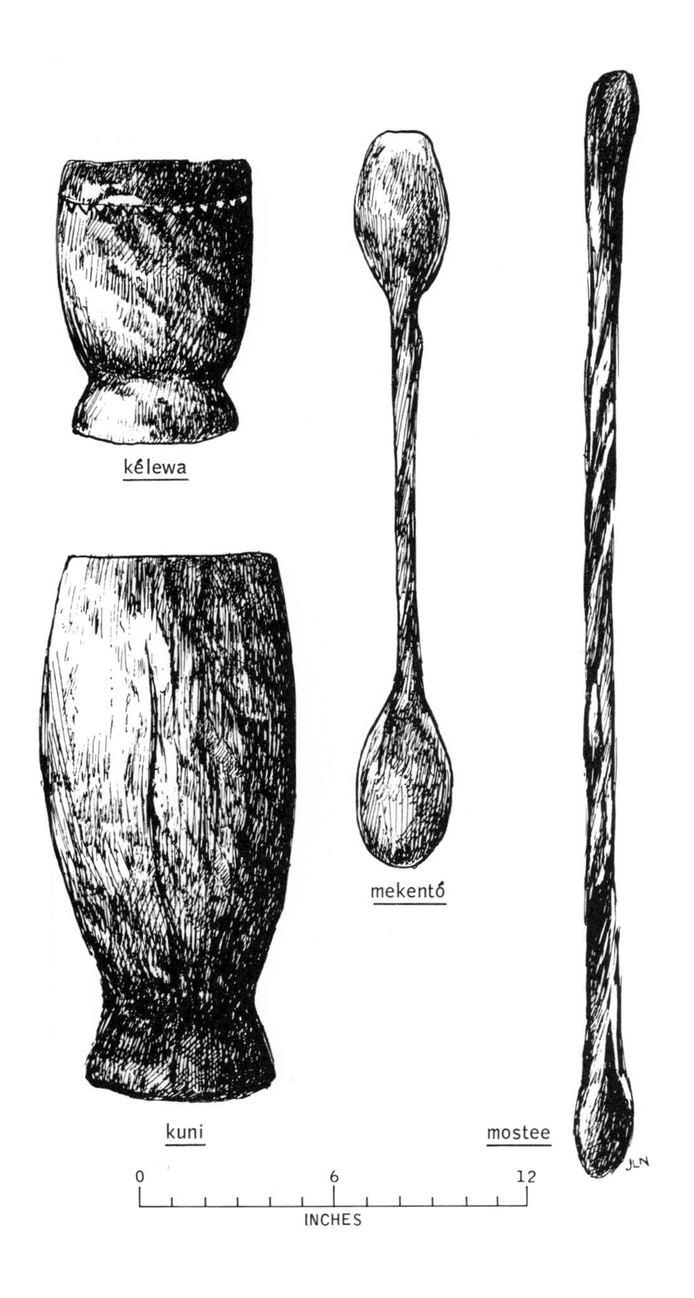
kêlewa
mekentó
kuni
mostee
JLN
0
6
12
INCHES

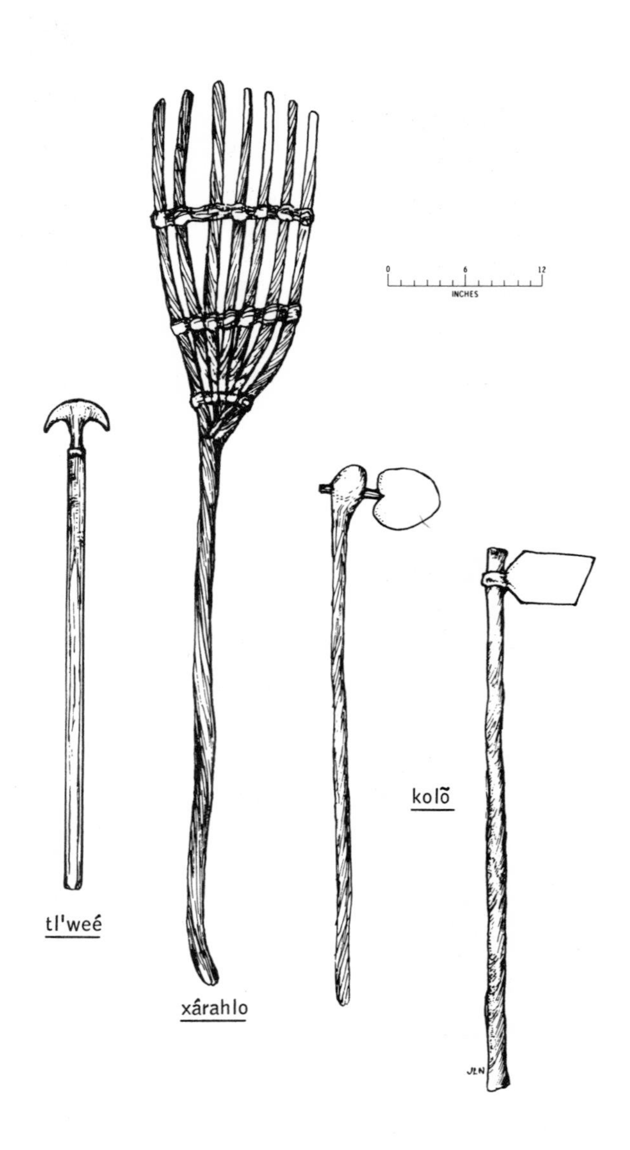
0 6 12
INCHES
tl'weé
xârahlo
kolõ
JLN

APPENDIX **B**

ANIMALS AND BIRDS HUNTED BY THE SANDAWE

APPENDIX B1 Animals Hunted for Food by the Sandawe

Common Name	Scientific Name[a]	Sandawe Name	Method of Hunting
Elephant	*Loxodonta africana*	*n/uaá*	Pit traps and poisoned arrows. Rarely hunted.
Black rhinoceros	*Diceros bicornis*	*tsõ*	Pit traps and poisoned arrows.
African buffalo	*Syncerus caffer*	*/eu*	Poisoned arrows and pit traps.
Masai giraffe	*Giraffa camelopardalis*	*tsámasu*	Poisoned arrows and pit traps.
Burchell's zebra	*Equus burchelli*	*dóro*	Poisoned arrows and pit traps.
Patterson's eland	*Taurotragus oryx*	*g!wá'a*	Poisoned arrows and pit traps.
Roan antelope	*Hippotragus equinus*	*sika*	Poisoned arrows and pit traps.
Sable	*Hippotragus niger*	*ébula*	Poisoned arrows and pit traps. Not found anymore.
Greater koodoo	*Tragelaphus stropsiceros*	*g!okomi* or *emau*	Poisoned arrows and pit traps.
Coke's hartebeest	*Alcelaphus buselaphus*	*korongoó*	Poisoned arrows and pit traps.
Wildebeest (gnu)	*Connochaetes taurinus*	*bús'*	Poisoned arrows and pit traps.
Impala	*Aepyceros melampus*	*paa*	Poisoned arrows. Very rare now.
Bushbuck	*Tragelaphus scriptus*	*tśa'wada*	Poisoned and unpoisoned arrows.
Grant's gazelle	*Gazella granti*	*tl'ak'ato*	Poisoned and unpoisoned arrows.

APPENDIX B1 *Continued*

Common Name	Scientific Name[a]	Sandawe Name	Method of Hunting
Thomson's gazelle	*Gazella thomsonii*	*!waká*	Unpoisoned arrows and nets.
Klipspringer	*Oreotragus oreotragus*	*!'echo*	Unpoisoned arrows and nets.
Bush duiker	*Sylvicarpa* sp.	*ts'indikó*	Nets and snare trap (*/'umuku*).
Kirk's dik-dik	*Rhynchotrogus kirkii*	*/híya*	Nets and snare trap (*/'umuku*).
Warthog	*Phacochoerus aethiopicus*	*//aá*	Poisoned arrows and pit traps.
Bush pig	*Potamochoerus choeropatamus*	*kéuto*	Poisoned and unpoisoned arrows plus pit traps.
Rock hyrax	*Dendrohyrax brucel*	*/weéna*	Harpoon-like poisoned arrows.
Spring hare	*Pedetes surdaster*	*dí'ra*	Nets and speared in holes.
Cape hare	*Lepus capensis*	*láá'e*	Unpoisoned arrows.
Porcupine	*Hystrix galeata*	*kialéé*	Unpoisoned arrows and snare trap (*/'umuku*).
White-tailed mongoose	*Ichneumia albicauda*	*nima*	Unpoisoned arrows and snare trap (*/'umuku*).
Banded mongoose	*Mungos mungo*	*xara*	Unpoisoned arrows and dogs.
Black-tipped mongoose	*Herpestes sanguineus*	*lóngos'*	Unpoisoned arrows and dogs.
Dwarf mongoose	*Helogale undulata*	*tsuaámaa*	Unpoisoned arrows and dogs.
Bush squirrel	*Paraxerus ochraceus*	*toómpéé*	Dogs and sticks used by small boys.
Ground squirrel	*Euxerus erythropus*	*kilíkili*	Dogs and sticks used by small boys.
Spectacled elephant shrew[b]	*Elephantulus rufescens*	*boló*	Rodent net and rock-fall trap (*makíba*).
Four-toed elephant shrew[b]	*Pterodromus tetradactylus*	*nangisé*	Rodent net and rock-fall trap (*makíba*).
Unstriped grass mouse[b]	*Arvicanthis abyssinicus*	*zakha'búr'*	Rodent net and rock-fall trap (*makíba*).
Spring mouse[b]	*Acomys selousi*	*méba'e*	Rodent net and rock-fall trap (*makíba*).
Honey badger	*Mellivora capensis*	*zírba*	Unpoisoned arrows.
Hedgehog	*Erinaceus pruneri*	*kamás'*	Sticks used by small boys.
Golden mole[b]	*Chlorotalpa* sp.	*kúta*	Unpoisoned arrows and speared in holes.
Aardvark	*Orycterupus afer*	*!'eenáá*	Speared in holes.
Lesser bushbaby	*Galago senegalensis*	*gáábéé*	Unpoisoned arrows.
Vervet monkey	*Ceropithecus aethiops*	*!haáku*	Unpoisoned arrows, dogs, and the basket-fall trap (*makíba*).

APPENDIX B1 *Continued*

Common Name	Scientific Name[a]	Sandawe Name	Method of Hunting
Yellow baboon	*Papio cynocephalus*	*//o//á*	Unpoisoned arrows.
Bat-eared fox	*Otocyon megalotis*	*búguli*	Nets and snare trap (*/'umuku*).
Black-backed jackal	*Canis mesomelas*	*monzó*	Nets and unpoisoned arrows.
Spotted hyena	*Crocuta crocuta*	*thékele*	Unpoisoned arrows. Very rarely hunted.
Brown hyena	*Hyaena brunnea*	*/'undu*	Unpoisoned arrows. Very rare.
Genet	*Genetta genetta*	*thandeé*	Snare trap (*/'umuku*) and unpoisoned arrows.
Civet	*Civettictis civetta*	*zorí*	Nets and unpoisoned arrows.
Serval	*Felis serval*	*ababúwaa*	Poisoned arrows and nets.
Cafer cat	*Felis ocreata*	*páka*	Unpoisoned and poisoned arrows. Very rare.
African wildcat	*Felis lybica*	*khàxà*	Poisoned arrows. Very rare.
Cheetah	*Acinonyx jubatus*	*babaaro*	Poisoned arrows. Rarely hunted.
Leopard	*Panthera pardus*	*théka*	Poisoned arrows. Rarely hunted.
Lion	*Panthera leo*	*//atsú*	Poisoned arrows. Very rarely hunted.

[a] Scientific terminology follows C. T. Astley Maberly, *Animals of East Africa* (Nairobi: D. H. Hawkins, Ltd., 1965).
[b] Identifications courtesy of College of African Wildlife, Mweka, Tanzania.

APPENDIX B2 Birds Hunted by the Sandawe[a]

Common Name	Scientific Name	Sandawe Name	Method of Hunting
Helmeted guinea fowl	*Numida mitrata*	*tsăk'e*	Arrows and the basket-fall trap (*makiba*). The outer tail and wing feathers are used for arrows.
Kenya crested guinea fowl	*Guttera pucherani*	*!wahli*	
Scaly francolin	*Francolinus squamatus*	*tlăká*	Arrows and both the nonreturn and basket-fall trap (*makíba*).
Grey-wing francolin	*Francolinus afer*	*xoweé*	
Coqui francolin	*Francolinus coqui*	*hundúrk'a*	

APPENDIX B2 *Continued*

Common Name	Scientific Name	Sandawe Name	Method of Hunting
Crested francolin	*Francolinus coqui*	*hundúrk'a*	
Yellow-billed hornbill	*Francolinus sephaena*	*mamatilíyo*	
Yellow-billed hornbill	*Tockus flavirostris*	*korímago*	Arrows only.
Red-billed hornbill	*Tockus erythrorphynchus*	*korímago*	
Vonder Kecken's hornbill	*Tockus deckeni*	*hokhorí*	
Ground hornbill	*Bucorvus leadbeateri*	*dūdū* or *dīdī*	
Donaldson Smith's nightjar	*Caprimulgus donaldsoni*	*!'umbaaro*	
Pennant-winged nightjar	*Semeiophorus vexillarius*	*!'umbaaro*	
African hoopoe	*Upupa africana*	*sambóó konkór'*	
Paradise flycatcher	*Terpsiphone viridis*	*dìints'saà*	
South African black flycatcher	*Melaenornis pammelaina*	*!hak'uk'u*	
Superb starling	*Spreo superbus*	*zilowé*	
Starling (?)		*sook'wa*	
Darnaud's barbet	*Trachyphonus darnaudii*	*!okorí!oko*	
Nubia woodpecker	*Campethera nubica*	*khēkhete*	
Swallow-tailed bee-eater	*Dicrocercus hirundineus*	*gil'*	
Malachite kingfisher	*Corythornis cristata*	*térerio*	
Black-bellied bustard	*Lissotis melanogaster*	*sōsóre*	
White-bellied bustard	*Eupodotis senegalensis*	*kwé kwe*	
Spotted morning warbler	*Chichladusa guttata*	*swérari*	
Yellow-bellied eremomela	*Eremomela icteropygialis*	*pí'no*	
Babler (?)	*Turdoides* sp.	*ts'emberk'u*	
Speckled pigeon	*Columba guinea*	*khongór'ma*	Arrows and bird-lime.
Red-eyed dove	*Streptopelia semitorquata*	*kōtsório*	
Ring-necked dove	*Streptopelia capicola*	*hekuku*	
Green pigeon	*Treron australis*	*!'wáá*	
Pink-breasted dove	*Streptopelia lugens*	*!waa*	
Emerald-spotted wood dove	*Turtur chalcospilos*	*!abelo*	
Namagua dove	*Oena capensis*	*swáá !abelo*	
Fisher's lovebird	*Agapornis fischeri*	*tl'ikík'i*	
Blue-naped mousebird	*Colius macrourus*	*píroór'*	
Cape rook	*Corvus capensis*	*alagwágwa*	
Southern double-collared sunbird	*Cinnyris chalybeus*	*ts'ents'éna*	
Variable sunbird	*Cinnyris venustus*	*ts'ents'éna*	
Scarlet-chested sunbird	*Chalcomitra senegalensis*	*ts'ents'éna*	
Amethyst sunbird	*Chalcomitra amethystina*	*ts'ents'éna*	
Marigua sunbird	*Cinnyris mariguensis*	*ts'ents'éna*	

APPENDIX B2 *Continued*

Common Name	Scientific Name	Sandawe Name	Method of Hunting
Golden-breasted bunting	*Emberizia flaviventris*	*dants'io*	
Lemon-rumped tinker-bird	*Pogoniulus leucolaima*	*bokío*	
African penduline tit	*Anthoscopus caroli*	*giróó*	
Fisher's straw-tailed whydah	*Vidua fischeri*	*gúle*	Arrows, birdlime, and the nonreturn trap (*thuundu*).
Paradise whydah	*Steganura paradisaea*	*gúle*	
Green-winged pytilia	*Pytilia melba*	*g/ing/oo*	
Jameson's firefinch	*Lagonosticta jamesoni*	*siísi*	
Red-billed firefinch	*Lagonosticta senegala*	*siísi*	
Cutthroat finch	*Amadina fasciata*	*siísi*	
Streaky seed eater	*Serinus striolatus*	*sã/'à*	
Bataleur	*Terathopius ecaudatus*	*póngo*	Arrows. These birds are not eaten; only the feathers are used.
Tawny eagle	*Aquila rapax*	*zíl'*	
Augue buzzard	*Buteo rufofuscus*	*k'waák'wà*	
Griffon vulture	*Gyps ruppellii*	*qwerésri*	
Ostrich	*Struthio camelus*	*sa'úta*	Poisoned and unpoisoned arrows. Eggs are also eaten.
Rufous-crowned roller	*Coracias naevia*	*wadó*	Birdlime only.
?	?	*!erek'ek'e*	
		kumpiná	
?	?	*dadawá*	Arrows only.
		ororía	
		tsho!wewe	
		zolówe	
		bahlíma	
		zékeke	
		tsuré	
		wakur'	
		wasénde	
		maliá	
		kobió	
?	?	*kúze*	Arrows and birdlime.
		k'wank'wári	
?	?	*tsuḷulu*	Birdlime and the nonreturn trap.

[a]Field and informant identifications from John G. Williams, *A Field Guide to the Birds of East and Central Africa* (London: Collins, 1963).

APPENDIX **C**

PLANTS USED BY THE SANDAWE

APPENDIX C Important Species of Natural Vegetation Used by the Sandawe[a]

Scientific Name	Sandawe Name	Use
Acacia kirkii	*samange*	Roots and bark boiled for stomach medicine.
Acacia millifera subsp. *detinens*	*rék'eto*	Wood for house construction.
Acacia nigrescens	*//'óxe*	Wood for house construction; bark for rope fibers.
Acacia nilotica subsp. *subalata*	*manange*	Roots boiled for stomach medicine and bark for "tea."
Acacia rovumae[b]	*ma'a*	Wood for house construction.
Acacia senegal	*k'án!a*	Wood for house construction.
Acacia tortilis subsp. *spirocarpa*	*áfa*	Bark for rope fibers.
Adansonia digitata[b]	*gelé*	Fruit eaten; seed pulp as fermenting agent in *pombe.* Bark for rope fibers and for hunting nets. Formerly source of bark cloth.
Afzelia quanzensis	*dō*	Wood for house construction, beehives, the grain mortar (*kuni*), and marimbas.
Albizia petersiana	*sán/a*	Wood for house construction.
Albizia tanganyicensis[b]	*maasin* (*maapin*)	Wood for beehives, drums, the carrying container (*la'sé*), the storage container for bows and arrows (*mú'a*), and beehive doors (*misiko sambala*).

APPENDIX C *Continued*

Scientific Name	Sandawe Name	Use
Allophylus rubifolius	*meráá*	Wood for building.
Aloe sp.	*!a!a'a*	Leaves boiled for diarrhea medicine.
Amaranthus graecizans	*moga*	Leaves boiled and eaten as a green vegetable; a Swahili *mchicha.*
Asparagus africanus	*konkór' !'intsha*	Roots boiled for gonorrhea medicine.
Aspilia sp.	*pangwé*	Roots boiled for children's stomach medicine and for menstrual cramps.
Berchemia discolor	*/'ók'óó*	Fruit eaten; wood for house construction; firewood, and the grain pestle (*mostee*).
Boscia angustifolia	*singísa*	Twigs for smoking milk before butter is made.
Boscia mossambicensis	*k'ats'awá*	Fruits eaten; wood for house construction.
Boscia salicifolia	*zimbau*	Wood for house construction.
Brachystegia microphylla	*k'aaya*	Bark used for rope fibers.
Brachystegia spiciformis	*inee*	Pods dried, boiled, and eaten; wood for house construction; bark for rope fibers and for food storage containers (*séngwá, kítongé, matundu*).
Bussea massaiensis subsp. *massaiensis*	*//'ánka*	Pods ground and boiled to be eaten as porridge; wood for house construction.
Canthium burtii	*namu (nam)*	Fruits eaten; wood for house construction, firewood, and the implement for vegetable chopping (*hlébera*).
Cassia abbreviata	*motóngolo*	Roots and bark boiled for stomach medicine; wood for house construction; pods used as candles.
Cassia singueana	*gelegéla*	Roots boiled for stomach medicine; wood for firewood; twigs for toothbrushes.
Cassipourea mollis	*!endi*	Wood for house construction.
Celosia schweinfurthiana	?	Young branches soaked in water for ear infection medicine.
Ceratotheca sesamoides	*betebeta*	Leaves boiled and eaten as green vegetable.
Ceropegia stenantha	*nongolo*	Roots boiled for children's stomach medicine.
Ceropegia sp.	*emperu*	Roots eaten, especially during famines.
Cissus integrifolia	*ts'ank'e songe*	Roots boiled for menstrual cramps.
Cissus trothae	*tuumbe*	Fruits eaten.
Cleome hirta	*nekeneka*	Roots and leaves boiled for measles medicine.
Clerodendrum ternatum	*goxoomo*	Roots boiled for medicine to counteract swelling of the extremities.

Appendix C *Continued*

Scientific Name	Sandawe Name	Use
Coccinia trilobata	*kóbá*	Leaves boiled and eaten, especially during famine.
Combretum apiculatum	*der'ma*	Wood for house construction.
Combretum collinum subsp. *taborense*	*konkohlawa*	Young branches chewed and liquid used as eye medicine.
Combretum langspicatum	*sóngalala*	Bark for rope fibers (for magic use only).
Combretum obovatum	*!axũ*	Roots boiled for stomach medicine.
Combretum schumanni	?	Wood for house construction.
Combretum trothae	*//ásna*	Wood for house construction and firewood.
Combretum zeyheri	*dérima*	Wood for house construction; branches for threshing sticks.
Combretum sp.	*//ho//ha*	Wood for firewood.
Commiphora caerulea	*/wému*	Wood for beehives, the piercing implements (*la'aá* and *rítimo*), and beehive doors (*misiko sambala*); young roots chewed for juices, especially during famine.
Commiphora eminii	*k'wana*	Sap used for fixing blades to ax handles.
Commiphora mollis	*/hin/ho*	Branches for smoking beehives; wood for the handles of the piercing implement (*lalaá*).
Commiphora pteleifolia	*/i/ida*	Roots boiled for headache medicine.
Commiphora swynnertonii	*ts'indiko*	Sap put on open wounds.
Commiphora ugogensis	*ámata*	Wood for beehives and the milking container (*kélewa*); sap for fixing blades to ax handles.
Corchorus trilocularis	*sagár'*	Leaves boiled and eaten as green vegetable.
Cordia ovalis	*tipã*	Fruits eaten.
Cordia rothii	*agweegwee*	Fruits eaten.
Crotalaria laburnifolia	*/e/édo*	Fruits eaten; wood for house construction.
Croton polytrichus	*gíndara'*	Bark of the root scraped to obtain "snuff" for headaches.
Cucumis aculeatus	*mumbu !hẽ*	Roots boiled for gonorrhea medicine.
Cynodon dactylon	*zwaárá*	Grass used on rooftops of houses and for construction of the small-bird trap (*thuundu*).
Cyphostemma knittelli	*tó'ondo*	Fruits eaten.
Dactyloctenium giganteum	*helá*	Seeds pounded and eaten, especially during famine.
Dalbergia arbutifolia	*támba*	Wood for ax handles, the handles of the honey container (*tómbóto*), house construction digging stick (*tl'weé*), and the handles of the wood-finishing implement (*xoronts'ima*).

APPENDIX C *Continued*

Scientific Name	Sandawe Name	Use
Dalbergia melanoxylon[b]	*ti'o*	Wood for house construction, combs, stools, and the grain pestle (*mostee*).
Delonix elata	*árange*	Seeds boiled and eaten.
Dichrostachys cinerea subsp. *nyassana*	*dégera*	Wood for house construction.
Diospyros cf. D. *fischeri*	*manats'úk'a*	Wood for house construction; roots for toothbrushes.
Dombea shupangae	*wáára*	Wood for bows, spears, fishing poles, musical bows, and house construction.
Ehretia obtusifolia	*limbo*	Roots boiled for spleen pains.
Entandophragma bussei	*ts'áawa*	Wood for beehives, the milking pot (*kélewa*), and the grain mortar (*kuní*).
Eragrostis aspera	*mozá*	Seeds mixed with bark from *maasin, n/wému,* or *ts'ik'àà* in water and rubbed on open sores.
Erianthemum commiphorae	*téga harabuta*	
Erythrina abyssinica	*tepetépa*	Bark boiled for menstrual cramps and chest pains.
Erythrocephalum setulosum	*biíri*	Roots boiled to obtain an eyewash.
Erythrococca atrovirens	*tarage*	Pods boiled and eaten.
Euphorbia bilocularis[b]	*áhleé*	Wood occasionally used for beehives; old and dried branches for night fires.
Euphoribia cuneata	*téga*	Sap for birdlime.
Euphorbia grantii	*ts'ík'àà*	Bark burned and placed on open wounds.
Euphorbia nyikae[b]	*dodó'ma*	Sap mixed with *!opa* for arrow poison.
Euphorbia quadrangulensis	*xen/á*	Sap mixed with water and used for stomach medicine.
Excoecaria bussei	*!opa*	Sap mixed with *dodó'ma* for arrow poison; wood for house construction.
Fagara chalybea	*k'ótso*	Wood for house construction; roots boiled for headache medicine.
Ficus fischeri	*láfa*	Fruits eaten; sap for birdlime.
Ficus petersii	*!'aku*	Fruits eaten; binding made from fibers of bark.
Ficus sycmorus	*sák'ana*	Fruits eaten.
Gerrardanthus lobatus	*póró*	
Grewia bicolor	*serekuúk'*	
Grewia holstii	*thenká*	

APPENDIX C *Continued*

Scientific Name	Sandawe Name	Use
Grewia mollis	*//hwaá*	Fruits eaten; wood for house construction, the storage container (*sánzo*), the field rake (*xárahlo*), the handles of the honey container (*tómbóto*), and the fish trap (*kibwábwa*).
Grewia platyclada	*xo'á*	Fruits eaten.
Grewia similis	*tsampure*	Fruits eaten.
Gyandropsis gynandra	*mazima*	Leaves cooked and eaten as a green vegetable.
Gymnema sylvestre	*mandábiri*	Roots boiled for medicine to combat swelling of the extremities.
Haplocoelum foliolosum	*/olok'oo*	Fruits eaten; wood for house construction.
Hibiscus lunarifolius	*kungee*	Branches for construction of fish traps.
Hippocratea buchananii	*weláá*	Wood for house construction and for the cooking spoon (*mekentó*).
Holarrhena fabrifugia	*n!an!ak'wa*	Roots boiled for gonorrhea medicine.
Hydnora johannis	*amamasòò*	Pulp eaten.
Hyphaene thebaica	*hagwé*	Fruits eaten and fronds used for weaving.
Indigofera lupatana	*la'ets'xunde*	Roots boiled for gonorrhea and pleurisy medicine.
Jatropha curcas	*bona*	Sap placed on open wounds.
Justicia salvioides	*hlanithéte*	Branches used for making shafts of arrows.
Kedrostis hirtella	*/'ĩko*	Fruits eaten.
Kigelia africana	*ráta*	Bark boiled for headache medicine.
Lanneà floccosa	*Kwílili*	Fruits eaten; bark used for rope fiber.
Lannea humulis	*g!omi*	
Lannea stuhlmannii	*/ámáka*	Fruits eaten; wood for house construction and the handle of the piercing implement (*lalaá*); bark boiled for menstrual cramps.
Launaea cornuta	*/eu genza*	Roots boiled for ankylostomiasis medicine.
Lightfottia abyssinica	*maxgo*	Roots boiled for measles medicine.
Lonchocarpus eriocalyx	*kúmani*	Wood for house construction.
Maerua edulis	*segeléé*	Fruits eaten.
Maerua tiphylla	*munuk'a*	Roots boiled for headache medicine.
Markhamia acuminata	*n!aama*	Wood for hoe and ax handles, firewood, and the cooking spoon (*mekentó*).
Markhamia obtusifolia	*hlohlobáá*	Wood for firewood and for making zithers.
Millettia paucijuga	*tl'abāthee*	Wood for house construction.
Momordica rostrata	*!'umphá*	Fruit boiled and eaten; root used as a substitute for soap.

APPENDIX C *Continued*

Scientific Name	Sandawe Name	Use
Monadenium schubei	*feege*	Roots boiled for stomach medicine.
Neorautenenia psuedopachyriza	*kwadaga*	Pods boiled and eaten.
Ochna ovata	*kwelegi*	Wood for house construction and the tip of the arrow (*ísoli*).
Ocimum americanum	*k'odaa*	Twigs used for constructing brooms.
Opilia celtidifolia	*ts'engere*	Fruits eaten.
Ostryoderris stuhlmanni	*lééba*	Wood for construction of beds, the grain mortar (*kuni*), and stools.
Panicum heterostachyum	*dwáro*	Roots eaten, especially during famine.
Pentas bussei	*bibi*	Branches used for making whistles.
Peponium vogelii	*hlampuka*	Fruits eaten.
Phyllanthus discoideus	*zíxaxa'*	Wood for house construction and firewood.
Phyllanthus engleri	*samángwe*	Roots boiled for menstrual cramps.
Portulaca pilosa	*kéri*	Roots chewed for toothache.
Pouzolzia parasitica	*dantl'a*	Young leaves boiled and eaten as a green vegetable.
Premna sp.	*parapapa*	Wood for house construction and hoe handles.
Premna sp.	*ts'ándu*	Wood for house construction and firewood.
Pseudolachnostylis maprouneifelia	*sesexu*	Wood for house construction; roots boiled for stomach medicine.
Pterocarpus angloensis	*!weéa*	Wood for house construction, beehives, and marimbas.
Rynchosia comosa	*!'indó*	Roots chewed for juices, especially during famine.
Rynchosia sp.	*alagwa !'indó*	Roots boiled or chewed for headache medicine.
Sacrostemma viminale	*lili'a*	Roots boiled for gonorrhea medicine.
Salvadora persica	*muléwa*	Fruits eaten plain or mixed with honey; branch for toothbrush.
Sanseviera sp.	*zaátsha*	Leaf used for rope fiber.
Schrebera oligantha	*pok'oo*	Branches chewed to obtain a liquid for eye infections.
Sclerocarya birrea	*án//uma*	Wood for construction of the grain mortar (*kuni*) and pestle (*mostee*), stools, and beehives; nuts are eaten.
Secamone stuhlmanii	*wakháma mogóngo*	Roots boiled for stomach medicine.
Securidaca longependunculata	*zatshia*	Roots boiled for headache medicine.
Sesamum angustifolium	*erénze*	Leaves boiled and eaten as green vegetables.

APPENDIX C *Continued*

Scientific Name	Sandawe Name	Use
Solanum incanum	*intór*	Roots boiled for stomach medicine.
Sporobolus pyramidalis	*kungu'e*	Roots chewed for stomach medicine.
Steganotaerina araliacea	*ferangomase*	Bark boiled for headache medicine; wood for hoe handles.
Sterculia africana	*tl'ágwa*	Bark for making fibers of binding twine.
Sterculia quinqueloba	*n!amahla*	Wood used for beehives.
Strophanthus eminii	*lobokhá* (*ferango*)	Young branches chewed as snakebite medicine; wood for handles of the fishing gaff (*mokhobé*), knife handles, and flutes.
Strychnos innocua	*g!éke*	Fruits eaten.
Strychnos potatorum	*tl'ikhaánka*	Wood for house construction.
Tamarindus indica	*/ank'á*	Pods ground up into flour for porridge.
Tapiphyllum floribundum	*sisímpiraa*	Fruits eaten.
Teclea glomerata	*motoxoro*	Roots boiled to obtain a medicine for kidney pains.
Tephrosia villosa	*tshiits'waa*	Roots boiled for headache and cold medicine.
Terminala sericea	*sen//a*	Wood for house construction.
Thylachium africanum	*mutungu*	Roots boiled and eaten, especially during famine.
Tinispora caffra	*zalã*	Sap placed on open wounds and sores.
Triaspis macropteron subsp. *massaiensis*	*g!okori*	Roots boiled for a medicine to counteract boils; wood for the handles of the honey container (*tómbóto*).
Turbina stenosiphon	*memeko*	Branches used to retain fire where matches are not available.
Turraea sp.	*//akástee*	Wood for house construction.
Vangueria acutiloba	*n!unk'máxae*	Fruits eaten.
Vangueria tomentosa	*n!uúk'*	Fruits eaten.
Vigna sp.	*ts'ik'sa*	Roots chewed for juices, especially during famine.
Vitex payos	*na'áso*	Fruits eaten; wood for house construction.
Vitex strickeri	*hláhlaxa*	Roots boiled for menstrual cramps; branches used for arrow shafts.
Waltheria indica	*khandágire*	Roots boiled for headache and cold medicine.
Withania somnifera	*xoxorik'a*	
Xeromphis taylori	*kipoloolo*	Fruits pounded and mixed with water for menstrual cramps.

APPENDIX C *Continued*

Scientific Name	Sandawe Name	Use
Ximenia americana	*//'aáya*	Fruits eaten.
Ximenia caffra	*!wadánda*	Fruits eaten.
Ziziphus mucronata	*ts'indi mak'o*	Fruits eaten; roots boiled for chest pain medicine.
Ascelpiadacea Family	*xaya*	Roots eaten, especially during famine.
?	*tangátshu*	Roots boiled and eaten, especially during famine.
?	*sánsa'mé*	Wood for handles of gaff fishing hook (*mokhobé*) and smoking pipes.

[a] Identifications from author's specimens by the East African Herbarium, Nairobi, Kenya.
[b] Field identifications by the author.

NOTES

CHAPTER ONE

1. Ackerman (1958), pp. 24–26.
2. Hägerstrand (1952), pp. 3–19.
3. Blaut (January 1959), p. 87.
4. Morgan and Moss (June 1965), p. 346.
5. Sahlins (1964), p. 134.
6. Frake (March 1962), p. 55.
7. Harris (1961).
8. Clayton (1964), p. 1.
9. Allan (1965), p. 470.
10. *Ibid.*

CHAPTER TWO

1. The "U" prefix designates "country of" in Swahili. Thus Usandawe means "country of the Sandawe." Most Sandawe now use this designation, since they have no comparable term in their own language. The form has also been incorporated into the English spoken in East Africa.
2. The first use of the designation Central Highlands is found in Werther (1898). Another term coined by the Germans for this same general area and referring to the fact that all drainage is interior is *die abflusslosen Gebiets.* See von Luschan, in Werther, *op. cit.*, pp. 323–81.

3. Gillman (1935), p. 32. The date comes from the text accompanying Geological Survey of Tanganyika, Quarter Degree Sheet 123, Kwa Mtoro (Dodoma: Geological Survey Division, 1961).
4. *Miombo* is a Swahili word which seems to encompass all the various species of *Brachystegia* trees. It does not refer to the total complex of the woodland, but has been widely used to do so by English speakers in eastern and southern Africa.
5. The present report follows the definition of colluvium given by Anderson (1957), p. 4, as material derived from the upper portion of a catena.
6. Trapnell and Langdale-Brown (1962), p. 99.
7. For discussions of the geology of this formation, see Rance (1961) and Quennell, McKinlay, and Aitken (1956), p. 242.
8. Burton (1860), p. 199.
9. Milne (1947), p. 219.
10. Burtt (1942), pp. 104–105.
11. Private communication with Douglas Turner, Senior Tsetse Officer, Ministry of Agriculture, Dar es Salaam, March 1966.
12. *Mbuga* is another Swahili word which has crept into widespread English usage in this part of Africa.
13. Gillman (1949), p. 18.
14. Coster (1960), p. 18.
15. All Sandawe words are written in the alphabet recommended by the International African Institute. For details, see International African Institute (1930) and Ward and Westermann (1933).
16. Rounce (1946).
17. Gillman (1949), pp. 18–19.
18. The matter of ethnic nomenclature is extremely confusing. For instance, the Hadza have been also called Kindiga, Tindiga, and Kangeju, plus several variations of these general forms; see Murdock (1959), p. 59. However, Dr. James Woodburn of the London School of Economics who studied the Hadza for several years maintains that Hadza is the recognized and preferred term among the people themselves (private communication, March 1965). A similar set of conflicting terms can be found for just about all the Central Highlands groups. The only way out of the predicament is to be arbitrary. I have attempted to find out what each of the peoples considers to be its proper name and then used this. Common alternatives will be given in parentheses. The widely employed Bantu-Wa prefix, meaning the group of people of the particular stem-name, e.g., Wasandawe, will be dropped as being redundant in English.
19. Coon (1963), p. 648, and Murdock (1959), p. 61.
20. Greenberg (1963), pp. 71–72.
21. The term Capoid is used by Coon (1963) p. 637, to refer to the Bushmen-Hottentot and their relations both living and dead. For Capoid affinities, see Murdock (1959), p. 60.
22. Trevor (1947), p. 76.
23. Bleek (November 1931), p. 427.
24. Tucker (1967), p. 679.
25. *Ibid.*
26. Bleek (July 1931), p. 274, and Huntingford (1953), p. 132.

27. Cole (1963), p. 335.
28. Greenberg, *op. cit.*, p. 49.
29. Nilo-Hamite is the traditional term and is most frequently encountered, but Greenberg, *ibid.*, pp. 85–86, objects to its use on the grounds of the confusion surrounding the term Hamite and thus proposes that Southern Nilote be substituted.
30. Beidelman (September 1960), p. 245.
31. Beidelman (March 1962), pp. 8–10.
32. Baumann (1894), p. 192.

CHAPTER THREE

1. Baumann, *op. cit.*, pp. 191–94.
2. von Luschan, *op. cit.*, pp. 328–35.
3. Reche (1914); Obst (1915 and 1923); Dempwolff (1916); Bagshawe (1924–25), pp. 219–27; and van de Kimmanade (1936), pp. 395–416.
4. Murdock (1959).
5. Private communication with Peter Rigby, Department of Sociology, Makerere University College, April 1966. Dr. Rigby is probably the world's leading authority on Gogo culture.
6. W. F. E. R. Tenraa, "*Nxako*, or the Sandawe Concept of War and Its Reflection on Tribal Structure" (unpublished manuscript in author's files, Oxford), p. 2.
7. All Sandawe literature presented in this report comes from the private collection of Mr. W. F. E. R. Tenraa. These and many others are scheduled for publication in the near future.
8. Baumann, *op. cit.*, p. 192.
9. Bagshawe, *op. cit.*, p. 225.
10. van de Kimmanade, *op. cit.*, p. 396.
11. Most items of material culture mentioned in the text and in the Appendixes are sketched in Appendix A. For others and for comparisons, the interested reader is referred to von Luschan, *op. cit.*, and Dempwolff, *op. cit.*
12. Private communication with John Kesby, Institute of Social Anthropology, Oxford University, September 1965. Mr. Kesby has recently completed three years of anthropological research among the Rangi.
13. von Sick (1915), p. 10. An unpublished English translation is available from Harold K. Schneider, Lawrence University, Appleton, Wisconsin.
14. *Ibid.*, and Harold K. Schneider, "Economics in an African Society: The Wahi Wanyaturu" (manuscript soon to be published, Lawrence University, Appleton, Wisconsin). Part already has been published; see Schneider (1964) and Schneider (1966).
15. Private communication with Dr. Peter Rigby, April 1966.
16. Beer is often used as the English equivalent, but this is probably an improper translation because of the lack of hops in *pombe* and an incomplete malting process. In making *pombe*, the grain is not dried or roasted after fermentation.
17. For an illustration of the Sandawe climbing rope, see Hunter (March 1952), p. 93.
18. The classification of the stingless bees is in a state of revision and consequently it is not possible to present the scientific designations.
19. van der Post (1961), pp. 58–71.

20. Bagshawe, *op. cit.,* p. 226.
21. Tanganyika Territory Ministry of Agriculture, "Central Province–Agricultural Policy 1954–55," Dar es Salaam, p. 20. (Mimeographed.)
22. Tenraa, *op. cit.,* p. 6.
23. Cory, "Gogo Law and Custom," Kampala, East African Institute of Social Research, 1951, Chap. VI, p. 1. (Typescript.); Cory, "Rimi Law and Custom," Kampala, East African Institute of Social Research, 1951, Chap. VI, p. 2. (Mimeographed). Rangi information from private communication with John Kesby, September 1965.
24. Cory, "Gogo Law and Custom," Chap. VI, p. 1, and Cory, "Rimi Law and Custom," Chap. VI, p. 5.
25. Schneider (1964), p. 85.
26. Cory, "Gogo Law and Custom," Chap. I, p. 4.
27. Schneider (1964), p. 153.
28. Cory, "Gogo Law and Custom," Chap. VII, pp. 6–7.
29. *Ibid.,* p. 1.
30. Schneider (1964), pp. 69–70.
31. Cory, "Gogo Law and Custom," Chap. VI, p. 8.
32. von Luschan, *op. cit.,* pp. 325–26, and Gray (March 1953), pp. 45–52.
33. Obst (1915), Bd. XXIX, p. 97.
34. Bagshawe, *op. cit.,* p. 224.
35. Private communication with Miss Marguerite Jellicoe, April 1966, former Community Development Officer, Singida District.
36. Private communication with W. F. E. R. Tenraa, May 1966.
37. Dempwolff, *op. cit.,* pp. 80–95.
38. Private communication with W. F. E. R. Tenraa, May 1966.
39. Bohannon (1963), pp. 337–39.
40. Schneider, *op. cit.,* p. 57.

CHAPTER FOUR

1. Hogbin (1958), p. 3.
2. Murdock (1956), pp. 249–50.
3. Locke and Stern (1946), p. 7.
4. Childe (1937), p. 20.
5. Allan, *op. cit.,* p. 273.
6. Baumann, *op. cit.,* pp. 189–92.
7. van de Kimmanade, *op. cit.,* p. 396.
8. The information on the descent of Turu clans comes from private communication with Miss Marguerite Jellicoe, April 1966.
9. Reche, *op. cit.,* p. 25.
10. von Luschan, *op. cit.,* p. 339.
11. An article on this subject will be forthcoming in the near future. Mr. Tenraa has kindly summarized some of the main points for me so that they could be included here.
12. Private communication with W. F. E. R. Tenraa, May 1966.
13. Tenraa, "*Nxako . . . ,*" p. 7.

14. Woodburn (September 1962), pp. 269–72.
15. *Ibid.*, p. 270.
16. James C. Woodburn, "The Social Organization of the Hadza of North Tanganyika" (unpublished PhD dissertation, Department of Anthropology, Cambridge University, 1964), p. 63.
17. Allan, *op. cit.*, pp. 261–66.
18. Wilson (September 1952), p. 35.
19. Hunter (March 1953), p. 31.

CHAPTER FIVE

1. Dempwolff, *op. cit.*, p. 72.
2. Figure from Kondoa District Administrative Book, now located in the National Archives, Dar es Salaam.
3. *Ibid.*
4. *Ibid.*
5. East African Statistical Department (1950), p. 12.
6. East African Statistical Department (1958), pp. 94–95.
7. For the full details on hunting regulations see United Republic of Tanzania (1965). (Mimeographed.)
8. Dempwolff, *op. cit.*, p. 79.
9. Tanganyika Territory Department of Agriculture (1945), p. 14.
10. East African Statistical Department (1953), p. 15.
11. Private communication with Thomas Augustino, January 1966, then field officer in Kwa Mtoro.
12. Disney and Haylock (January 1956), p. 142.

CHAPTER SIX

1. Fuggles-Couchman (1964), p. 18.
2. Hadley E. Smith (1965), pp. 10–11.
3. Good (1964), p. 332.
4. See Penman (1963), and Linacre (September 1963), pp. 165–77.
5. Dagg (January 1965), p. 297.
6. For evaluations of the Penman estimate, see Wang and Wang (December 1962), pp. 582–88; Jesuitas, Aglibut, and Sandoval (September 1961), pp. 165–80, and Tanner and Pelton (October 1960), pp. 3391–3413.
7. Penman (July 1948), pp. 120–45, and Penman (1963), pp. 40–43.
8. McCulloch (January 1965), pp. 286–95.
9. Manning (1956), pp. 472–73.
10. Thornthwaite (January 1948), pp. 75–77.
11. Curry (April 1962), p. 176.
12. For an example of maize, see Dagg, *op. cit.*, pp. 298–99. His figures were not used in this paper because the growing season at Maguga in Kenya is different from that in Usandawe.
13. Grimes and Clarke (October 1962), pp. 74–80.

14. Cobley (1956), p. 26.
15. Evans (April 1955), p. 265.
16. Ruthenberg (1964), pp. 170–77.
17. Jones (1965), p. 35.
18. Private communication with Gordon Anderson, Research Chemist, Northern Research Centre, Tengeru, Tanzania, March 1966.
19. Coster, *op. cit.,*pp. 26–31.
20. Allan, *op. cit.,* p. 37.
21. Milne, *op. cit.,* pp. 221–22.
22. Anderson (1963), p. 9.
23. O'Hoore (1963). (Mimeographed.)
24. Anderson, *op. cit.*
25. The general information on human disease in Usandawe was supplied through private communication with the Area Medical Officer, Kondoa, October 1965.
26. Private communication with the Iambi Leprosarium, October 1965.
27. Desert Locust Control Organization for Eastern Africa (1965), and Joyce (1961).
28. Allan, *op. cit.*, p. 48.
29. Data on plant pests and diseases supplied by Arthur Hammersly, Assistant Director of Agriculture, Ministry of Agriculture, Forests and Wildlife, Dar es Salaam, March 1966.

CHAPTER SEVEN

1. Hutchison (1965), p. 46.
2. Heady (1960), pp. 27–28 and end map.
3. *Ibid.*, p. 86.
4. Information courtesy of the Veterinary division of the Ministry of Agriculture, Dar es Salaam, November 1965 and March 1966.
5. Mackenzie and Simpson (1964), p. 69.
6. Jackson (December 1933), p. 211.
7. *Ibid.*, p. 270.
8. Based on the testimony of local informants and on private communication with Douglas Turner, Senior Tsetse Officer, Ministry of Agriculture, March 1966.
9. Swynnerton (November 1936), p. 57.
10. Jackson, *op. cit.*, p. 220.
11. Private communication with Douglas Turner, March 1966.
12. For a detailed discussion of the methods of tsetse control, see Buxton (1955), pp. 489–590.
13. For the latest information, see Buxton (above) and Glasgow (1963).

CHAPTER EIGHT

1. Huxley (1961), p. 55.
2. Darling (1960), pp. 23–25.
3. Huxley, *op. cit.*, pp. 34–35.

4. Talbot *et al.* (1965), pp. 2, 16, 20–23. For further details, consult the excellent bibliography in this publication, especially the entries under L. M. Talbot and G. A. Petrides.
5. For a discussion of rock paintings in Tanzania, see The Tanganyika Society (July 1950), pp. 1–61.
6. Darling, *op. cit.*, p. 95.
7. Dale and Greenway (1961), p. 362.
8. *Ibid.*, pp. 96, 212, 589.
9. Watt and Breyer-Brandwijk (1962).
10. Smith (1960), p. 72.
11. *Ibid.*, pp. 45–56.
12. For a discussion of the costs of various beekeeping operations, the reader is referred to Smith, pp. 83–91.
13. Identification of the fish species comes as a courtesy of Dr. R. G. Bailey, Fisheries Research Officer, Ministry of Agriculture, Nyegezi, Tanzania.
14. Hickling (1961), pp. 90–104.
15. I suspect that the reason why nets were not used in fishing (although they were used in hunting) is that none of the local fibers appears to be sufficiently water-resistant.

CHAPTER NINE

1. Recent attempts at measuring biomass production might be the answer to the problem of establishing objective absolute standards. For the theory, see Margalef (November–December 1963), pp. 357–75, and for application see Ovington, Heitkamp, and Lawrence (Winter 1963), pp. 52–63.
2. Allan, *op. cit.*
3. Conklin (1959), p. 63, and Carneiro (1960), pp. 230–31.
4. Allan, *op. cit.*, pp. 20–37, 49–65.
5. *Ibid.*, p. 24.
6. *Ibid.*, p. 23.
7. *Ibid.*, p. 26.
8. Hunter (May 1966), pp. 151–54.

CHAPTER TEN

1. Fried (1962), pp. 315–16.
2. Huxley, *op. cit.*, p. 26.
3. Webster and Wilson (1966), pp. 196–97.
4. Ruthenberg, *op. cit.*, pp. 86–87.
5. van Rensburg (April 1955), pp. 251–53.
6. Dumont (1966).

REFERENCES

BOOKS

Ackerman, Edward A. *Geography as a Fundamental Research Discipline*. (University of Chicago, Department of Geography Research Paper No. 53.) Chicago: University of Chicago Press, 1958.

Allan, William. *The African Husbandman*. Edinburgh: Oliver & Boyd, 1965.

Anderson, B. *A Survey of Soils in the Kongwa and Nachingwea Districts of Tanganyika.* Reading, England: Department of Agricultural Chemistry, Faculty of Agriculture and Horticulture, 1957.

Baumann, Oscar. *Durch Massailand zur Nilquelle.* Berlin: Dietrich Reimer, 1894.

Bohannan, Paul. *Social Anthropology*. New York: Holt, Rinehart and Winston, Inc., 1963.

Burton, R. T. *The Lake Regions of Central Africa*. New York: Harper & Bros., 1860

Buxton, Patrick A. *The Natural History of Tsetse Flies.* (London School of Hygiene and Tropical Medicine Memoir 10.) London: H. K. Lewis & Co., Ltd., 1955.

Clayton, Eric S. *Agrarian Development in Peasant Economies.* Oxford: Pergamon Press, 1964.

Cobley, Leslie S. *An Introduction to the Botany of Tropical Crops.* London: Longmans, Green and Co., Ltd., 1956.

Cole, Sonia. *The Prehistory of East Africa.* New York: Macmillan Co., 1963.

Coon, Carleton S. *The Origin of Races.* New York: Alfred A. Knopf, 1963.

Coster, F. M. *Underground Water in Tanganyika.* Dar es Salaam: Government Printer, 1960.

Dale, Ivan R., and Greenway, P. J. *Kenya Trees and Shrubs.* Nairobi: Buchanan's Kenya Estates, Ltd., 1961.

Darling, F. Fraser. *Wildlife in an African Territory.* London: Oxford University Press, 1960.

Dumont, René. *False Start in Africa.* Trans. by Phyllis Nauts Ott. London: Andre Deutsch, 1966.

Fuggles-Couchman, N. R. *Agricultural Change in Tanganyika: 1945–1960.* Stanford, California: Food Research Institute, Stanford University, 1964.

Glasgow, J. P. *The Distribution and Abundance of Tsetse.* (International Series of Monographs on Pure and Applied Biology, Zoology Division, V. 20.) Oxford: Pergamon Press, 1963.

Good, Ronald. *The Geography of the Flowering Plants.* 3d ed., revised. London: Longmans, 1964.

Harris, J. F. *Summary of the Geology of Tanganyika, Part IV: Economic Geology.* Dar es Salaam: Government Printer, 1961.

Heady, Harold F. *Range Management in East Africa.* Nairobi: Government Printer, 1960.

Hickling, C. F. *Tropical Inland Fisheries.* London: Longmans, 1961.

Hogbin, H. Ian. *Social Change.* London: C. A. Watts & Co., Ltd., 1958.

Huxley, Julian. *The Conservation of Wild Life and Natural Habitats in Central and East Africa.* Paris: UNESCO, 1961.

International African Institute. *Practical Orthography of African Languages.* 2d ed., revised. London: Oxford University Press, 1930.

Locke, Alain, and Stern, Bernhard J. *When Peoples Meet.* New York: Hinds, Hayden & Eldredge, Inc., 1946.

Mackenzie, P. Z., and Simpson, R. M. *The African Veterinary Handbook.* 4th ed., revised. Nairobi: Sir Isaac Pitman & Sons, Ltd., 1964.

Murdock, George Peter. *Africa, Its Peoples and Their Culture History.* New York: McGraw-Hill Book Co., Inc., 1959.

Penman, H. L. *Vegetation and Hydrology.* (Commonwealth Bureau of Soils, Technical Communication No. 53.) Farnham Royal, Bucks, England: Commonwealth Agricultural Bureaux, 1963.

Post, Laurens van der. *The Heart of the Hunter.* New York: Morrow, 1961.

Quennell, A. M., McKinlay, A. C. M., and Aitken, W. G. *Summary of the Geology of Tanganyika, Vol. I: Introduction and Stratigraphy.* Dar es Salaam: Government Printer, 1956.

Rounce, Norman V. *The Agriculture of the Cultivation Steppe of the Lake, Central and Western Provinces.* Salisbury, Rhodesia: Longmans, Green and Co., 1946.

Ruthenberg, Hans. *Agricultural Development in Tanganyika.* (Ifo-Institut Für Wirtschaftsforschung, Afrika-Studien, 2.) Berlin: Springer-Verlag, 1964.

Smith, Francis G. *Beekeeping in the Tropics.* London: Longmans, 1960.

Talbot, Lee M., and others. *The Meat Production Potential of Wild Animals in Africa.* (Commonwealth Bureau of Animal Breeding and Genetics, Technical Communication No. 16.) Farnham Royal, Bucks, England: Commonwealth Agricultural Bureaux, 1965.

Tanganyika Territory Department of Agriculture. *Agriculture in Tanganyika.* Dar es Salaam: Government Printer, 1945.

Ward, Ida C., and Westermann, D. *Practical Phonetics for Students of African Languages.* London: Oxford University Press, 1933.

Watt, J. M., and Breyer-Brandwijk, M. G. *Medicinal and Poisonous Plants of Southern and Eastern Africa.* 2d ed., revised. London: E. & S. Livingstone, Ltd., 1962.

Webster, C. C., and Wilson, P. N. *Agriculture in the Tropics.* London: Longmans, 1966.

Werther, C. Waldemar. *Die mittleren Höchlander des nordlichen Deutsch-Ost-Afrika.* Berlin: Herman Paetel, 1898.

Williams, John G. *A Field Guide to the Birds of East and Central Africa.* London: Collins, 1963.

ARTICLES

Bagshawe, F. J. "The Peoples of the Happy Valley (East Africa), Part III, The Sandawe," *Journal of the African Society,* XXIV (1924–25), pp. 219–27.

Beidelman, T. O. "The Baraguyu," *Tanganyika Notes and Records,* LV (September 1960), pp. 245–78.

Beidelman, T. O. "A Demographic Map of the Baraguyu," *Tanganyika Notes and Records,* LVIII (March 1962), pp. 8–10.

Blaut, James M. "Microgeographic Sampling," *Economic Geography,* XXXV (January 1959), pp. 79–88.

Bleek, D. F. "The Hadzapi or Watindiga of Tanganyika Territory," *Africa,* IV (July 1931), pp. 273–86.

Bleek, D. F. "Traces of Former Bushman Occupation in Tanganyika," *South African Journal of Science,* XXVIII (November 1931), pp. 424–29.

Burtt, B. D. "Some East African Vegetation Communities." Edited by C. H. N. Jackson, *Journal of Ecology,* XXX (February 1942), pp. 65–146.

Carneiro, Robert L. "Slash-and-Burn Agriculture: A Closer Look at Its Implications for Settlement Patterns," in *Selected Papers of the Fifth International Congress of Anthropological and Ethnological Sciences, September 1–9, 1956.* Edited by Anthony F. C. Wallace. Philadelphia: University of Pennsylvania Press, 1960, pp. 229–34.

Childe, V. Gordon. "A Prehistorian's Interpretation of Diffusion," in *Independence, Convergence and Borrowing in Institutions, Thought and Art,* Harvard Tercentenary Publications. Cambridge: Harvard University Press, 1937, pp. 3–21.

Conklin, Harold C. "Population–Land Balance Under Systems of Tropical Forest Agriculture," *Proceedings of the Ninth Pacific Science Congress, 1957,* VII. Bangkok: Secretariat, Ninth Pacific Science Congress, 1959, p. 63.

Curry, Leslie. "The Climatic Resources of Intensive Grassland Farming: The Waikato, New Zealand," *Geographical Review,* LII (April 1962), pp. 174–94.

Dagg, Mathew. "A Rational Approach to the Selection of Crops for Areas of Marginal Rainfall in East Africa," *East African Agricultural and Forestry Journal,* XXX (January 1965), pp. 296–300.

Dempwolff, Otto. "Die Sandawe," *Abhandlungen Hamburgischen Kolonialinstitute,* Band XIX. Hamburg: L. Friederechsen & Co., 1916.

Disney, H. J. deS., and Haylock, J. W. "The Distribution and Breeding Behaviour of the Sudan Dioch (*Quela Q. Aethiopica*) in Tanganyika," *East African Agricultural Journal,* XXI (January 1956), pp. 141–47.

Evans, A. C. "A Study of Crop Production in Relation to Rainfall Reliability," *East African Agricultural Journal,* XX (April 1955), pp. 263–67.

Frake, C. O. "Cultural Ecology and Ethnography," *American Anthropologists,* LXIV (March 1962), pp. 53–59.

Fried, Morton H. " Land Tenure, Geography and Ecology in the Contact of Culture," *Readings in Cultural Geography.* Edited by Philip L. Wagner and Marvin W. Mikesell. Chicago: University of Chicago Press, 1962, pp. 302–17.

Gillman, Clement. "A Vegetation-types Map of Tanganyika Territory," *Geographical Review,* XXXIX (January 1949), pp. 7-37.

Gray, Robert F. "Notes on Irangi Houses," *Tanganyika Notes and Records,* XXXV (March 1953), pp. 45–52.

Greenberg, Joseph H. "The Languages of Africa," *International Journal of American Linguistics,* Part II, XXIX (1963).

Grimes, R. C., and Clarke, R. T. "Continuous Arable Cropping with the Use of Fertilizers," *East African Agricultural and Forestry Journal,* XXVII (October 1962), pp. 74–80.

Gulliver, P. H. "A Tribal Map of Tanganyika," *Tanganyika Notes and Records,* LII (March 1959), pp. 61–74.

Hägerstrand, Torsten. "The Propagation of Innovation Waves," *Lund Studies in Geography, Series B, Human Geography,* IV (1952), pp. 3–19.

Hunter, G. "A Sandawe Climbing Rope," *Tanganyika Notes and Records,* XXXIII (March 1952), p. 93.

Hunter, G. "Hidden Drums in Singida District," *Tanganyika Notes and Records,* XXXIV (March 1953), pp. 28–32.

Hunter, John M. "Ascertaining Population Carrying Capacity Under Traditional Systems of Agriculture in Developing Countries," *The Professional Geographer,* XVII (May 1966), pp. 151–54.

Huntingford, G. W. B. "The Southern Nilo-Hamites," *Ethnographic Survey of Africa, Part VIII, East Central Africa.* London: International African Institute, 1953.

Hutchison, H. G. "The Undeveloped Potential of Livestock," in *Agricultural Development in Tanzania.* Edited by Hadley E. Smith. (University College, Institute of Public Administration, Study No. 2.) Dar es Salaam: Oxford University Press, 1965, pp. 46–49.

Jackson, C. H. N. "On an Advance of Tsetse Fly in Central Tanganyika," *Transactions of the Royal Entomological Society of London,* LXXXI (December 1933), pp. 205–221.

Jackson, C. H. N. "On Two Advances of Tsetse Fly in Central Tanganyika," *Proceedings of the Royal Entomological Society of London,* Series A, XXV (1950), pp. 29–32.

Jesuitas, E. F., Aglibut, A. P., and Sandoval, A. R. "Measurement of Potential Evapotranspiration as a Basis for Irrigating Crops," *Philippine Agriculturist,* XLV (September 1961), pp. 165–80.

Jones, William O. "Environment, Technical Knowledge and Economic Development," in *Ecology and Economic Development in Tropical Africa.* Edited by David Brokensha. (International Studies Research Series No. 9.) Berkeley: University of California Press, 1965, pp. 29–48.

Kimmanade, Martin van de. "Les Sandawe," *Anthropos,* XXXI (1936), pp. 395–416.

Linacre, E. T. "Determining Evapotranspiration Rates," *Journal of the Australian Institute of Agricultural Science,* XXIX (September 1963), pp. 165–177.

Luschan, F. von. "Beitrage zur Ethnographie des abflusslosen Gebiets von Deutsch-Ost-Afrika," in *Die mittleren Hochländer des nördlichen Deutsch-Ost-Afrika.* Edited by C. Waldemar Werther. Berlin: Hermann Paetel, 1898, pp. 323–81.

McCulloch, J. S. G. "Tables for the Rapid Computation of the Penman Estimate of Evaporation," *East African Agricultural and Forestry Journal,* XXX (January 1965), pp. 286–95.

Manning, H. L. "The Statistical Assessment of Rainfall Probability and Its Application in Uganda Agriculture," *Empire Cotton Growing Corporation, Research Memoirs,* XXIII (1956), pp. 460–80.

Margalef, R. "On Certain Unifying Principles in Ecology," *The American Naturalist,* XCVII (November–December 1963), pp. 357–74.

Milne, G. "A Soil Reconnaissance Journey Through Parts of Tanganyika Territory, December 1935 to February 1936." Edited by Clement Gillman. *Journal of Ecology,* XXXV (December 1947), pp. 192–265.

Morgan, W. B., and Moss, R. P. "Geography and Ecology: The Concept of Community and Its Relation to Environment," *Annals of the Association of American Geographers,* LV (June 1965), pp. 339–50.

Murdock, George Peter. "How Cultures Change," in *Man, Culture and Society.* Edited by Harry L. Shapiro. New York: Oxford University Press, 1956, pp. 247–60.

Obst, Erich. "Das Abflusslose Rumpfschollenland im nordöstlichen Deutsch-Ostafrika," *Metteilungen des Geographische Gesellschaft in Hamburg,* Bands XXIX & XXXV. Hamburg: L. Friederechsen & Co., 1915 and 1923.

Ovington, J. D., Heitkamp, D., and Lawrence, D. B. "Plant Biomass Productivity of Prairie, Savanna, Oakwood and Maize Fields in Central Minnesota," *Ecology,* XLIV (Winter 1963), pp. 52–63.

Penman, H. L. "Natural Evaporation from Open Water, Bare Soil and Grass," *Proceedings of the Royal Society of London (A),* CLXLIII (July 1948), pp. 120–45.

Reche, Otto. "Zur Ethnographic des Abflusslosen Gebietes Deutsch Ostafrikas," *Abhandlungen Hamburgischen Kolonialinstitute,* Band XVII. Hamburg: L. Friederechsen & Co., 1914.

Rensburg, H. J. van. "Land Useage in Semi-Arid Parts of Tanganyika," *East African Agricultural Journal,* XX (April 1955), pp. 247–53.

Sahlins, Marshall O. "Culture and Environment: The Study of Cultural Ecology," in *Horizons of Anthropology.* Edited by Sol Tax. Chicago: Aldine Publishing Company, 1964, pp. 132–47.

Schneider, Harold K. "Economics in East African Aboriginal Societies," in *Economic Transition in Africa.* Edited by Melville J. Herskovits and Mitchell Harwitz. Evanston, Ill.: Northwestern University Press, 1964, pp. 53–76.

Schneider, Harold K. "Turu Ecology: Habitat, Mode of Production and Society," *Africa,* XXXVI (July 1966), pp. 254–68.

Sick, Eberhard von. "Die Waniaturu (Walimi)," *Baessler-Archiv,* V, No. 1 (1915), pp. 1–62.

Smith, Hadley E. "Agriculture and the Five-Year Plan of Tanganyika," in *Agricultural Development in Tanzania.* Edited by Hadley E. Smith. (University College, Institute of Public Administration Study No. 2.) Dar es Salaam: Oxford University Press, 1965, pp. 1–23.

Swynnerton, C. F. M. "The Tsetse Flies of East Africa," *Transactions of the Royal Entomological Society of London,* LXXXIV (November 1936), pp. 1–579.

Tanner, C. B., and Pelton, W. L. "Potential Evapotranspiration Estimates by the Approximate Energy Balance Method of Penman," *Journal of Geophysical Research,* LXV (October 1960), pp. 33403–33407.

The Tanganyika Society. "Tanganyika Rock Paintings, A Guide and Record," *Tanganyika Notes and Records,* XXIX (July 1950), pp. 1–61.

Thornthwaite, C. W. "An Approach Toward a Rational Classification of Climate," *Geographical Review,* XXXVIII (January 1948), pp. 85–94.

Trapnell, Colin G., and Langdale-Brown, Ian. "The Natural Vegetation of East Africa," in *The Natural Resources of East Africa.* Edited by E. W. Russell. Nairobi: D. A. Hawkins, Ltd., 1962, pp. 92–102.

Trevor, J. C. "The Physical Characters of the Sandawe," *Journal of the Royal Anthropological Institute,* LXXVII, Part 1 (1947), pp. 61–78.

Tucker, A. N. "Fringe Cushitic, An Experiment in Typological Classification," *Bulletin of the School of Oriental and African Studies,* XXIX (October 1967), pp. 655–79.

Wang, Jen Yu, and Wang, S. C. "Simple Graphical Approach to Penman's Method for Evaporation Estimates," *Journal of Applied Meteorology,* I (December 1962), pp. 582–88.

Wilson, Gordon McL. "The Tatoga of Tanganyika," *Tanganyika Notes and Records,* XXXIII (September 1952), pp. 35–47.

Woodburn, James. "The Future of the Tindiga," *Tanganyika Notes and Records,* LIX (September 1962), pp. 268–73.

PUBLIC DOCUMENTS

Anderson, B. "Soils of Tanganyika," *Ministry of Agriculture Bulletin No. 16.* Dar es Salaam: Government Printer, 1963.

Desert Locust Control Organization for Eastern Africa. *Third Annual Report of the Directors.* Nairobi: The English Press, Ltd., 1965.

East African Statistical Department. *African Population of Tanganyika Territory. Geographical and Tribal Studies.* Nairobi: E.A.S.D., 1950.

East African Statistical Department. *Report on the Analysis of the Sample Census of African Agriculture 1950 Tanganyika.* Nairobi: E.A.S.D., 1953.

East African Statistical Department. *General African Census August 1957, Tribal Analysis Part II, Territorial Census Areas.* Dar es Salaam: Government Printer, 1958.

Gillman, Clement. "Geomorphological Notes. Usandawe," *Annual Report, Tanganyika Territory Geological Survey, 1934.* Dar es Salaam: Government Printer, 1935, pp. 31–32.

Joyce, R. J. V. *Report of the Desert Locust Survey 1st June, 1955–31st May, 1961.* East African Common Services Organization. Nairobi: Government Printer, 1961.

United Republic of Tanzania. *The Fauna Conservation Ordinance.* Dar es Salaam: Government Printer, 1965.

UNPUBLISHED MATERIAL

Cory, Hans. "Gogo Law and Custom," Kampala, East African Institute of Social Research, 1951. (Typescript.)

Cory, Hans. "Rimi Law and Custom," Kampala, East African Institute of Social Research, 1951. (Typescript.)

O'Horre, J. "Soils Map of Africa at a Scale of 1/5,000,000." Revised definitions, Leopoldville, CCTA Symposium, 1963. (Mimeographed.)

Rance, Hugh. "The Kilimatinde Cement." Unpublished Master's thesis, Department of Geology, University of Natal, 1961.

Schneider, Harold K. "Economics in an African Society: The Wahi Wanyaturu." Unpublished manuscript, Lawrence University, Appleton, Wisconsin.

Tanganyika Territory Ministry of Agriculture. "Central Province–Agricultural Policy 1954–55," Dar es Salaam. (Mimeographed.)

Tanzania National Tourist Board. "Notes on Hunting and Game Photography in the Republic of Tanzania," Dar es Salaam, 1965. (Mimeographed.)

Tenraa, W. F. E. R. "Nxako, or the Sandawe Concept of War and Its Reflection on Tribal Structure." Unpublished manuscript, Oxford.

Woodburn, James C. "The Social Organization of the Hadza of North Tanganyika." Unpublished Ph.D. dissertation, Department of Anthropology, Cambridge University, 1964.

OTHER SOURCES

Geological Survey of Tanganyika. Quarter Degree Sheet 124, Kelema. Dodoma, 1960.

Geological Survey of Tanganyika. Quarter Degree Sheet 123, Kwa Mtoro. Dodoma, 1961.

Geological Survey of Tanganyika. Quarter Degree Sheet 143, Meia Meia. Dodoma, 1963.

Survey Division, Tanganyika Department of Lands and Surveys. Physiographical Map of Tanganyika. *Atlas of Tanganyika*. 3d ed., revised. Dar es Salaam, 1950, p. 1.

Survey Division, Tanganyika Department of Lands and Surveys. 1:50,000 Sheets 103/IV, 104/111, 123/1–IV, 124/1–111, 142/1 & 11, 143/1. Dar es Salaam.

Survey Division, Tanzania Ministry of Lands, Settlement and Water. Tanzania, 1:2,000,000. Dar es Salaam, 1965.